52

Springer Proceedings in Physics

Managing Editor: H. K. V. Lotsch

30 *Short-Wavelength Lasers and Their Applications* Editor: C. Yamanaka

31 *Quantum String Theory*
Editors: N. Kawamoto and T. Kugo

32 *Universalities in Condensed Matter*
Editors: R. Jullien, L. Peliti, R. Rammal, and N. Boccara

33 *Computer Simulation Studies in Condensed Matter Physics: Recent Developments*
Editors: D. P. Landau, K. K. Mon, and H.-B. Schüttler

34 *Amorphous and Crystalline Silicon Carbide and Related Materials*
Editors: G. L. Harris and C. Y.-W. Yang

35 *Polycrystalline Semiconductors: Grain Boundaries and Interfaces*
Editors: H. J. Möller, H. P. Strunk, and J. H. Werner

36 *Nonlinear Optics of Organics and Semiconductors*
Editor: T. Kobayashi

37 *Dynamics of Disordered Materials*
Editors: D. Richter, A J. Dianoux, W. Petry, and J. Teixeira

38 *Electroluminescence*
Editors: S. Shionoya and H. Kobayashi

39 *Disorder and Nonlinearity*
Editors: A. R. Bishop, D. K. Campbell, and S. Pnevmatikos

40 *Static and Dynamic Properties of Liquids*
Editors: M. Davidović and A. K. Soper

41 *Quantum Optics V*
Editors: J. D. Harvey and D. F. Walls

42 *Molecular Basis of Polymer Networks*
Editors: A. Baumgärtner and C. E. Picot

43 *Amorphous and Crystalline Silicon Carbide II: Recent Developments*
Editors: M. M. Rahman, C. Y.-W. Yang, and G. L. Harris

44 *Optical Fiber Sensors*
Editors: H. J. Arditty, J. P. Dakin, and R. Th. Kersten

45 *Computer Simulation Studies in Condensed Matter Physics II: New Directions*
Editors: D. P. Landau, K. K. Mon, and H.-B. Schüttler

46 *Cellular Automata and Modeling of Complex Physical Systems*
Editors: P. Manneville, N. Boccara, G. Y. Vichniac, and R. Bidaux

47 *Number Theory and Physics*
Editors: J.-M. Luck, P. Moussa, and M. Waldschmidt

48 *Many-Atom Interactions in Solids*
Editors: R. M. Nieminen, M. J. Puska, and M. J. Manninen

49 *Ultrafast Phenomena in Spectroscopy*
Editors: E. Klose and B. Wilhelmi

50 *Magnetic Properties of Low-Dimensional Systems II: New Developments*
Editors: L. M. Falicov, F. Mejía-Lira, and J. L. Morán-López

51 *The Physics and Chemistry of Organic Superconductors*
Editor: G. Saito and S. Kagoshima

52 *Dynamics and Patterns in Complex Fluids: New Aspects of the Physics-Chemistry Interface*
Editors: A. Onuki and K. Kawasaki

Volumes 1 – 29 are listed on the back inside cover

Dynamics and Patterns in Complex Fluids

New Aspects of the Physics-Chemistry Interface

Proceedings of the
4th Nishinomiya-Yukawa Memorial Symposium,
Nishinomiya City, Japan, October 26–27, 1989

Editors: A. Onuki and K. Kawasaki

With 125 Figures

Springer-Verlag
Berlin Heidelberg New York London
Paris Tokyo Hong Kong Barcelona

Professor Dr. Akira Onuki
Research Institute for Fundamental Physics, Kyoto University, Kyoto 606, Japan

Professor Dr. Kyozi Kawasaki
Department of Physics, Kyushu University, Fukuoka 812, Japan

ISBN-13: 978-3-642-76010-5 e-ISBN-13: 978-3-642-76008-2
DOI: 10.1007/ 978-3-642-76008-2

Softcover reprint of the hardcover 1st edition 1990

2154/3140 – 5 4 3 2 1 0 – Printed on acid-free paper

Preface

The fourth Nishinomiya-Yukawa Memorial Symposium, devoted to the topic of dynamics and patterns in complex fluids, was held on October 26 and 27, 1989, in Nishinomiya City, Japan, where ten invited speakers gave their lectures. A one-day meeting, comprising short talks and poster sessions, was then held on the same topic on October 28 at the Research Institute for Fundamental Physics, Kyoto University. The present volume contains the 10 invited papers and 38 contributed papers presented at these two meetings.

The symposium was sponsored by Nishinomiya City, where Prof. Hideki Yukawa once lived and where he wrote the celebrated paper describing the work that was later honored by a Nobel prize. The topic of the fourth symposium was chosen from one of the most vigorously evolving and highly interdisciplinary fields in condensed matter physics. The field of complex fluids is very diverse and still in its infancy and, as a result, the definition of a complex fluid varies greatly from one researcher to the next. One of the objectives of the symposium was to clarify its definition by explicitly posing a number of potentially rich problems waiting to be explored. Indeed, experimentalists are disclosing a variety of intriguing dynamical phenomena in complex systems such as polymers, liquid crystals, gels, colloids, and surfactant systems. We, the organizers, hope that the symposium will contribute to the increasing importance of the field in the coming years.

We have been impressed lately by the growing richness of phenomena, particularly those at the boundaries between physics, chemistry, and engineering. A number of important topics in current statistical physics have stemmed from such interdisciplinary fields of science. They are related, for example, to defects, mesoscopic ordering, patterns, nonlinear response, random interfaces, etc. We believe that the symposium succeeded in revealing a wealth of new problems lying in such interface areas.

We would like to thank the ten invited lecturers, who are well known as pioneers, and all the enthusiastic participants, numbering roughly 160. The symposium was made possible by the generous support of Nishinomiya City. We express our sincere gratitude to Mr. Yoneji Yagi, mayor of Nishinomiya City, and Prof. Ziro Maki, chairman of the steering committee of the Nishinomiya symposia.

Kyoto, Fukuoka
March 1990

Akira Onuki
Kyozi Kawasaki

Contents

Part I

Membranes and Surfactant Systems

Adhesion and Roughness of Biological Model Membranes

W. Helfrich[1,2] *and B. Klösgen*[1]

[1]Fachbereich Physik, Freie Universität Berlin,
Arnimallee 14, D-1000 Berlin 33, Germany
[2]Institute for Theoretical Physics, University of California,
Santa Barbara, CA 93106, USA

Abstract. Several electrically neutral biological model membranes have been found to separate in plenty of water, adhering to each other only when under lateral tension. The large contact angles ($\geq 40°$) of induced adhesion suggest a submicroscopic roughness of the membranes which at zero tension absorbs at least 10 times as much area as do undulations. A model of cooperative local saddles has been proposed to explain the roughness. We review the situation, present an electron micrograph which possibly shows a pattern of saddles, argue that saddles may be destroyed by too much tension and then need a long time to reappear, and list other evidence for anomalous behavior of these membranes.

1. Introduction

Fluid amphiphilic bilayers and monolayers perform out-of-plane fluctuations which roughen the membrane surface. The only out-of-plane fluctuations of fluid membranes established to date are thermal undulations. They are controlled by the bending rigidity and the lateral tension of the membrane. Thermal undulations can be very large in the absence of lateral tension. Flaccid vesicles of the stiffest fluid membranes known, the biological model membranes, still display well visible undulations under a phase contrast microscope.

Some fits of theoretical shapes to experimental contours of rotationally symmetric vesicles [1] indicated that the bending elastic energy as a function of the principal curvatures obeys the usual quadratic form including a linear term for spontaneous curvature. Recent studies of separate undulation modes permitting multiple measurement of the bending rigidity of a single membrane seem to confirm this. The same rigidity was found for all the modes regardless of their wavelength [2].

Theory predicts the ensemble of thermal undulations to affect some material properties of fluid membranes. For instance, they should absorb real membrane area (*i.e.* mass) most of which can be recovered by applying lateral tensions [3-6]. Undulations will also produce a repulsive force between parallel membranes as one membrane hinders the undulations of the other [5-7]. Moreover, the undulations should reduce the effective bending rigidity that refers to a non-fluctuating mean surface

Springer Proceedings in Physics, Vol. 52 **Dynamics and Patterns in Complex Fluids**
Editors: A. Onuki · K. Kawasaki

rather than the fluctuating real surface of the membrane [8-10]. Like the absorbed area the decrement of the rigidity increases logarithmically with membrane size. Beyond a certain size, which is expressed by a persistence length, fluid membranes are expected to crumple [11].

The actual undulation forces of very flexible membranes whose other interactions could be neglected were found to agree within experimental error with a theoretical formula [12,13]. Moreover, the area absorbed by undulations has been measured for some biological model membranes as a function of lateral tension. The latter was produced by sucking vesicles into a micropipette [14] or by exposing them to an electric field [15]. The absorbed area showed the expected logarithmic decrease with increasing tension so that the prefactor of the logarithm can be used to derive the bending rigidity. Finally, the persistence length and effective rigidity have proved to be useful concepts in the understanding of bicontinuous microemulsion and related systems [16,17]. Measurements of these quantities are difficult and have not yet been reported.

Despite the impressive evidence for a well-behaved bending rigidity and for undulations being the only out-of-plane fluctuations, there are other data which are not in accordance with this picture. They were obtained exclusively with some electrically neutral biological model membranes, namely the bilayers of lecithins (=phosphatidylcholines or PC's), cephalins (=phosphatidylethanolamines or PE's), and a natural digalactosyldiacylglycerol (=DGDG). The first two materials and the last one abound in animal and vegetable membranes, respectively, so that the bilayers and their vesicles may indeed be regarded as simple models of biological membranes and cells. The chemical and physical properties of all the membranes have been studied in many ways and by many authors.

The need to postulate for these membranes a submicroscopic roughness much larger than that of undulations emerged in the course of investigations in our group of their mutual adhesion. Several years ago it was found that egg and other lecithin membranes separate indefinitely if the lipid is left to swell in plenty of water or NaCl solution [18]. Later on, rare adhesive contacts were noticed which seemed to be induced by associated lateral tensions [5]. Newer data appear to satisfy two scaling laws suggested by undulation theory which involve lateral tension, adhesion energy, and the equilibrium mean spacing of induced adhesion [19,20]. Initially this looked like a confirmation of the action of undulations [21], but then it turned out that induced adhesion is about 100 times more sensitive to lateral tension than is to be expected [19,20]. The postulated roughness is needed to bridge the gap. Very similar discrepancies were subsequently found with dimyristoyl PC, some PE's, and a natural DGDG in pure water [22]. As the only exception, DGDG membranes display spontaneous mutual adhesion, but they do so only in NaCl solution and at low temperatures [23].

In the following we outline the experimental data on adhesion induced by lateral tension. Subsequently, we point out the glaring contradictions concerning the numerical factors of the two scaling laws which strongly suggest a membrane roughness much larger than that of undulations. We then propose a model of the postulated new roughness. Its basic element is a local saddle deformation of very high principal curvatures which in a planar membrane must be surrounded by two highs and two lows. Being cooperative, the conjectured saddle structures might be able to warp the bilayers. The model has been described elsewhere [24]. Here we elaborate on possible consequences of an energy barrier for the formation of saddles. Also for the first time, we show a photograph of an egg lecithin vesicle which differs from most others by a graininess of its membrane. It was obtained by transmission electron microscopy of a frozen sample [25]. Size and packing of the grains are consistent with a rather dense system of saddle structures. In the conclusion we list other indications of anomalous behavior of electrically neutral biological model membranes.

2. Induced Adhesion: Experimental Results

Most of the observations of adhesion induced by lateral tension were made with egg lecithin swollen in pure water. We describe them in some detail before briefly considering other systems. Rare accidental adhesive contacts were utilized in one study [15], while in another induced adhesion was brought about reproducibly by cooling [20].

2.1 Accidental Adhesion

In these experiments a very small quantity ($\leq 0.1\%$) of the lipid was spread on the object slide and the rest of the sample cell was filled with water. Within a few hours most of the egg lecithin tended to transform into vesicular structures largely consisting of single lamellae. The visibly fluctuating vesicles did in general not adhere to each other. However, rare adhesive contacts were found in some samples. Two examples are given in Figs. 1 and 2. Note that the membrane is seen where it is parallel to the optical axis of the phase contrast microscope and in the plane of focus. Figure 1 shows the symmetric adhesion of two single membranes which is characterized by a flat contact area and equal contact angles of both bilayers. The contact angle is measured between the asymptotic direction of the full membrane and the alignment of the membrane where it is bound. The rather local rounding of the membrane before it merges into the contact area is due to the bending rigidity which opposes an abrupt change of direction. Finding the asymptotic direction is difficult or impossible if other deformations of the membrane dominate. Figure

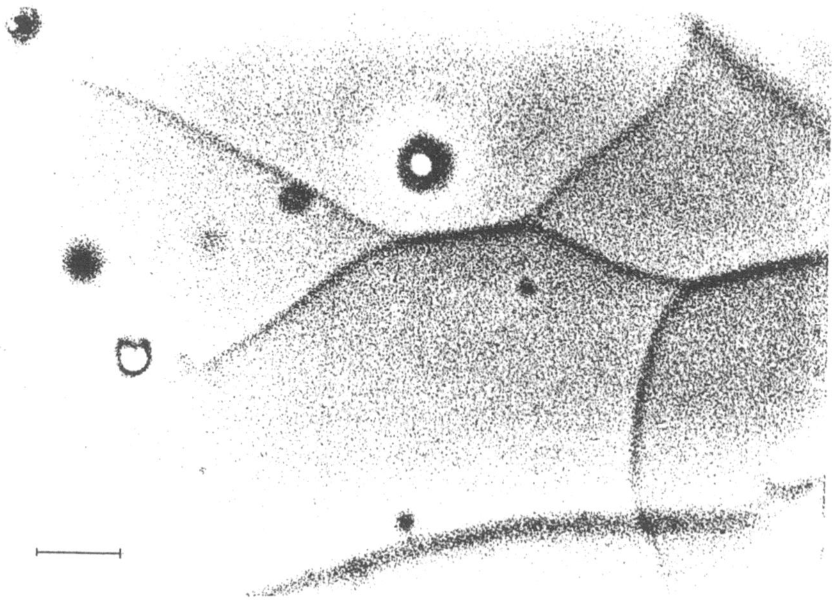

1. Adhesive contact of two single egg lecithin membranes. The contact (upper middle) is practically symmetric, all four contacts being near 40°. Note the well visible contact roundings next to the contact area. The contact angles refer to the "asymptotic" directions a few μm away. Some irregularities are fluctuations because of extremely low tensions. The bar represents 10μm.

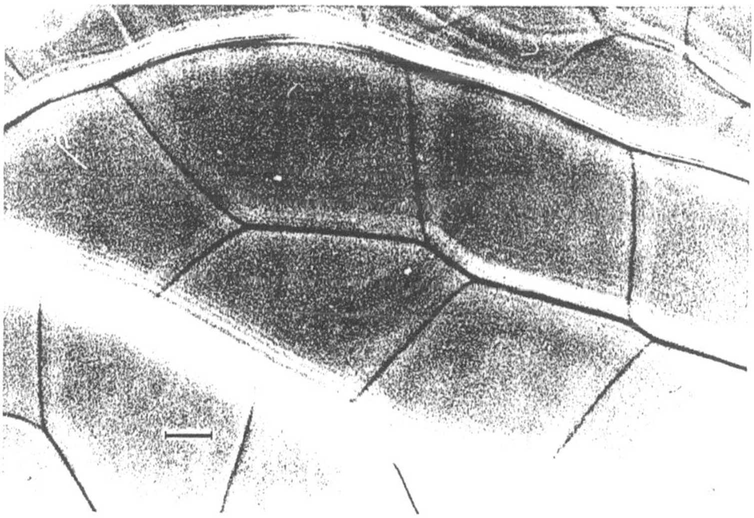

2. A branched structure of egg lecithin membranes displaying two symmetric adhesions. In one case, further membranes join one by one to form a thicker and thicker bundle. Contact roundings are shorter and fluctuations weaker than in Fig. 1. The bar represents 10μm.

2 displays a branched structure containing two symmetric adhesions of single bilayers and the adhesion of additional bilayers to a bundle composed of an increasing number of membranes. For the most part contact roundings are well defined.

What can be learned from a large number of pictures of this sort? One can read contact angles and determine or estimate the rounding length ξ_r, *i.e.* the orientational correlation length of the membrane, which obeys [5,19-22]

$$\xi_r^2 = \kappa/\sigma. \tag{1}$$

If the bending rigidity κ is known one can then calculate from ξ_r the lateral tension σ. Employing the Young equation

$$g_a = 2(1 - \cos\psi)\sigma, \tag{2}$$

where ψ is the contact angle, one can also derive g_a, the adhesion energy per unit area. The factor 2 enters for symmetric contacts, the only ones whose data were collected in accidental adhesion. The tensions calculated on the basis of $\kappa = 1 \cdot 10^{-12}$erg were in the range of $3 \cdot 10^{-6}$ to $3 \cdot 10^{-4}$dyn cm^{-1}. Adhesive contacts with lower tensions did not occur, those with higher tensions could not be evaluated because ξ_r was below the limit of optical resolution.

The contact angle in the range of tensions covered was $(40 \pm 5)°$. It showed some scatter but no significant dependence on lateral tension. In conjunction with the Young equation (2), a constant contact angle implies that the adhesion energy is proportional to the lateral tension. For $\psi = 40°$ one has

$$g_a = 0.6\sigma. \tag{3}$$

A double logarithmic plot of g_a as calculated from (2) versus σ as calculated from (1) is given in Fig. 3. It collects all the experimental data on symmetric adhesion of egg lecithin bilayers.

2.2 Adhesion Brought About By Cooling

Lateral tension and the ensuing adhesion can be generated in a controlled and reversible manner by cooling well-aligned multilayer systems swollen at elevated temperatures. Systematic experiments of this type were performed only with egg lecithin [20]. Let us first summarize the swelling of these systems which had been studied previously at room temperature [26]. The membranes of the slightly hydrated liquid were aligned largely parallel to the glass slides by the well-known method of squeezing, the final cell thickness being typically $20\mu m$. The remaining 90% of the cell volume was filled with pure water. From the beginning, the planar phase of the multilayer system had a semicylindrical border eliminating membrane edges. The early stages of swelling were charac-

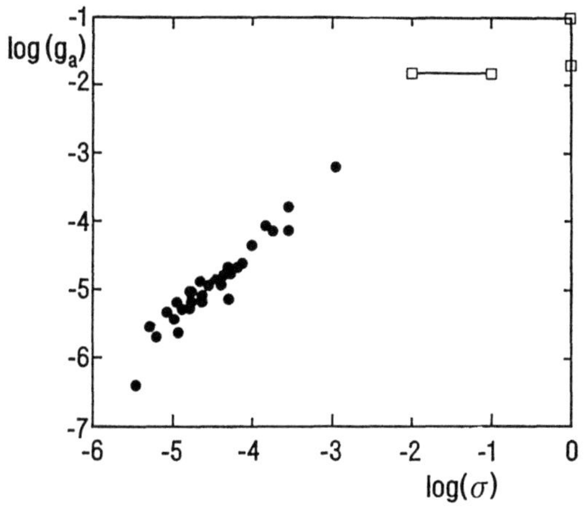

3. Adhesion energy per unit area, g_a, plotted versus lateral tension σ. The plot contains all the data on symmetric adhesion of egg lecithin. The point at 10^{-3}dyn cm^{-1} was obtained with dimyristoyl PC. The horizontal line represents measurements of E. Evans [33]. The squares arbitrarily placed at 1 dyn cm^{-1} stand for the results of Rand *et al.* [39] and Marra and Israelachvili [28] obtained with liposomes and immobilized membranes, respectively.

terized by a rapid growth of myelin cylinders from the border into the free water. The myelin cylinders spanned the thickness of the cell and appeared to be folded border. Their growth finally stopped, but the increase of the mean membrane spacing continued, reaching 20 nm in the planar phase after *ca.* 2 weeks at room temperature. The mean spacing in the planar phase was obtained from the intensity of egg lecithin fluorescence. After some swelling, the spacing tended to be *ca.* 2 times larger in the semicylindrical border and in the myelin cylinders, as is sketched in Fig. 4.

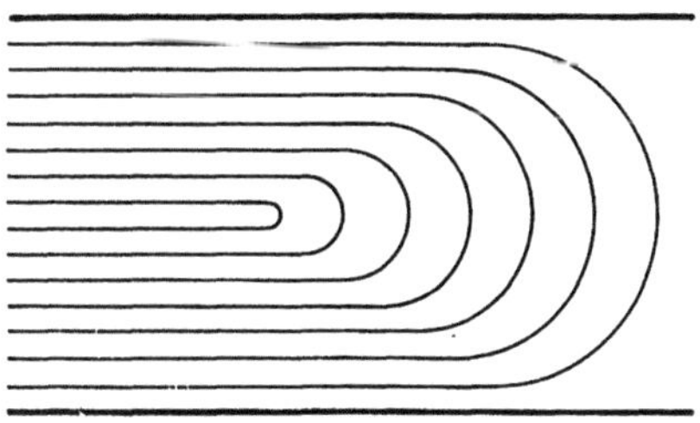

4. Schematic cross section through multilayer system from top to bottom of the sample cell. The real number of membranes is around 1000. The mean membrane spacing is less in the planar phase (left) than in the semicylindrical border (right), in accordance with observation. The sample is viewed from the top.

In the new experiments [20] swelling was done at elevated temperatures of typically 50 to 60° C. Subsequent cooling led to phase separations in the form of isolated water-rich regions. If the temperature drop and the mean spacing of the membranes in the planar phase were large enough, one could see single membranes running between stacks of membranes which apparently were regions of adhesion. The phenomena were in general restricted to the curved regions and reversible by reheating. Once single membranes and contact rounding were discernible, the rounding length and thus the tension σ did not noticeably change with further cooling. The mean spacing in the regions of adhesion, which should be the equilibrium spacing, also seemed constant and was about equal to the spacing in the planar phase.

These effects were interpreted in terms of a fairly quantitative model. The membranes which are mostly in the planar phase shrink upon cooling [27]. The shrinkage forces water into the curved regions which are, in fact, seen to widen. Being able to diminish in length, the latter have a capacity for water which is proportional to the radius of each membrane (or, more precisely, intermembrane space). The variable capacity requires a slight adaption of the mean membrane spacings in the planar phase as a function of this radius. Both decrease and increase of the mean spacings are easy if the tension is large enough to reduce the total force between membranes to zero. As a result, the tension will not rise beyond this value. The self-limiting tension reaches this value only asymptotically since some net force on each membrane must remain to balance the pressure difference across the membrane. The latter is produced by the action of the same tension in the semicylindrical border.

Measurable quantities in these experiments were the mean spacing in the planar phase as well as contact angle and rounding of single membranes (and bundles) running from one stack of membranes in adhesion to another. The contact angle of single membranes was about 70°. Inserting this into Eq. (2) after cancelling the factor 2 on the right hand side leads to

$$g_a = 0.5\sigma. \tag{4}$$

Computing the tension from the rounding length and taking the equilibrium mean spacing of induced adhesion, $\bar{z}_{eq}$, to be equal to the calculated mean spacing in the planar phase, one can plot σ versus $\bar{z}_{eq}$. The result is shown in Fig. 5 which collects all the data, indicating the (large) experimental error and the ranges covered by $\bar{z}_{eq}$ and σ. The data points suggest the linear relationship

$$\bar{z}_{eq}^2 \approx \frac{1}{400}\frac{kT}{\sigma}, \tag{5}$$

especially when 1 nm is added to the membrane in thickness on both of its sides to account for the so-called hydration forces [28].

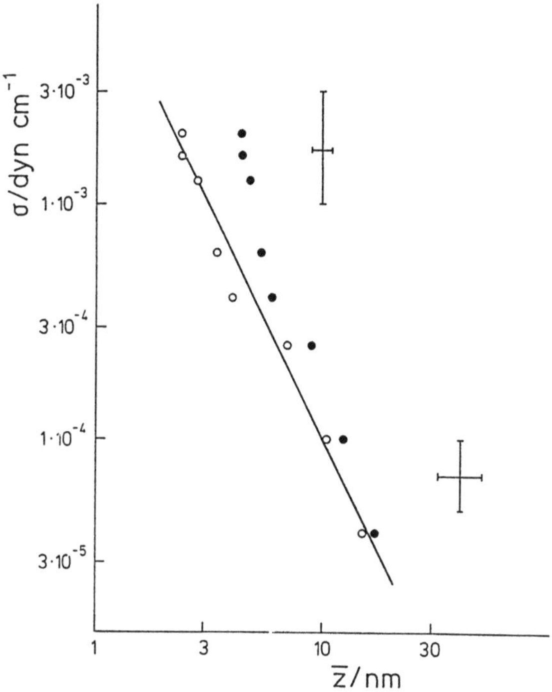

5. Squared equilibrium mean spacing of induced adhesion, $\bar{z}_{eq}$, plotted versus lateral tension. The solid dots are the raw data, the open dots were obtained after subtracting 2 nm from the spacing (see text). The straight line represents $\bar{z}_{eq}^2 \sim 1/\sigma$.

2.3 Other Systems

Accidental adhesion was also studied for dimyristoyl lecithin (DMPC) in pure water and egg lecithin in 30 mM NaCl solution. The data could not be distinguished from those obtained with egg lecithin in pure water. Higher salt concentrations slowed the swelling of egg lecithin, and DMPC developed no suitable structures in salt solutions. The accidental adhesive contacts of the bilayers of some PE's and DGDG in pure water were similar to those of egg lecithin [22]. The contact angles with these materials were larger than with the lecithins but remained less than 90°. However, contact angles of 90° were common below the unbinding transition of DGDG membranes in 0.1 M NaCl solution [23].

3. Induced Adhesions: Analysis

The two apparent scaling laws of adhesion induced by lateral tension

$$g_a \sim \sigma \quad \text{and} \quad \bar{z}_{eq}^2 \sim 1/\sigma$$

can be derived from undulation theory if the "direct" force between adjacent membranes is expressed by the half-space approximation of van der Waals attraction

$$f_{\mathrm{vdw}} = -\frac{H}{6\pi z^3}, \qquad (6)$$

where H is the Hamaker constant. This approach ignores short-range hydration forces and the weakening of van der Waals attraction at large spacings ($\bar{z} \gtrsim 20$ nm [29]) due to the finite bilayer thickness of *ca.* 4 nm. In the absence of any direct force there are two correlation lengths for membranes fluctuating in a stack (or between rigid walls). One is the rounding length ξ_r (1) which may also be viewed as an orientational correlation length, and the other is the lateral correlation length $\xi_{\|}$ which obeys

$$\xi_{\|}^2 \approx \frac{\kappa}{kT}\,\bar{z}^2$$

if a numerical factor near unity is omitted. It is rather obvious that lateral tension should reduce pure undulation forces by a factor F which is a function only of the ratio

$$\xi_r^2/\xi_{\|}^2 = \frac{\sigma\bar{z}^2}{kT}$$

where κ cancels out. For the stack this implies [5,7]

$$f_{\mathrm{und}} = \frac{3\pi^2}{64}\frac{(kT)^2}{\kappa\bar{z}^3}F\left(\frac{\sigma\bar{z}^2}{kT}\right),$$

with $F = 1$ for $\sigma = 0$, and a very similar result should hold for the pair. In the absence of spontaneous adhesion, *i.e.*, roughly for $H/6\pi < 3\pi^2(kT)^2/64\kappa$, one can now try to calculate $\bar{z}_{eq}$ from the equilibrium condition in the form

$$f_{\mathrm{und}}(\sigma) + f_{\mathrm{vdw}} = 0,$$

putting $z = \bar{z}$. This leads directly to the scaling law $\bar{z}_{eq}^2 \sim 1/\sigma$ whose numerical factor may be obtained by cutting the membrane into pieces of the right size to be treated as the particles of a one-dimensional ideal gas [5,20]. The approach is naive, presupposing the validity of the superposition principle for direct and undulatory forces which need not be correct [30]. However, such reservations do not apply to the simple scaling law. A more detailed discussion of the scaling law is given elsewhere [31]. Combining $\bar{z}_{eq}^2 \sim 1/\sigma$ with the van der Waals interaction energy per unit area in the half-space approximation,

$$g_{\mathrm{vdw}} = -\frac{H}{12\pi z^2},$$

immediately suggests the other scaling law which was observed, $g_a \sim \sigma$. Inspection shows that its validity also requires $g_f \sim \sigma$, where g_f is the stretching energy of the free membrane [20,31]. The latter proportionality is indeed valid (see below).

The apparent confirmation of undulation theory by the two scaling laws gives way to failure when the numerical factors are included. Let us show this for $g_a \sim \sigma$, considering symmetric adhesion of two single bilayers. The adhesion energy may be defined by the equation

$$g_a = 2(g_f - g_b)$$

where g_f and g_b are the energies of stretching (per unit area) of the free and the bound membrane, respectively. Not only g_a and g_f but also g_b must be positive for $\sigma > 0$ because $g_b < 0$ would mean spontaneous adhesion. (This is seen best by regarding g_b as a function of the strain rather than the stress.) Accordingly, one may write

$$g_a < 2g_f \, .$$

The energy of stretching of the free membrane has been calculated to be [3-6]

$$g_f = \frac{kT}{8\pi\kappa}\,\sigma + \frac{1}{2\lambda}\,\sigma^2 .$$

The second term contains the stretching modulus, measured by Kwok and Evans to be $\lambda = 140$ dyn cm^{-1} [27] and is negligible at the tensions in question. The second term results from the flattening of the undulations of an unstretchable fluid membrane. With $\kappa = 1 \cdot 10^{-12}$ erg we thus obtain

$$g_a < 3.2 \cdot 10^{-3}\,\sigma. \tag{7}$$

This theoretical result differs by more than two orders of magnitude from the relationship (3) which was obtained from the experiments. The glaring contradiction cannot be eliminated by allowing for the large spread of measured bending rigidities, about a factor of two, and the increase of projected area through the flattening of undulations which can be no more than a few percent [3-6].

It has therefore been argued that the enormous gap between (3) and (7) can only be closed if a much larger reservoir of absorbed area, at least 30% of the projected area, is pulled flat by the very low tensions of the experiments. A considerable roughness highly sensitive to lateral tension could also explain why the squared equilibrium spacings of induced adhesion are about a hundred times smaller than undulation theory and approximate measurements of the undulation forces at $\sigma = 0$ seem to permit [20,31]. The postulated new roughness has to be on a small scale of no more than a few 100 nm as it is invisible under the light microscope. On the other hand, its structures should not be much smaller than this so that tensions of 10^{-5} dyn cm^{-1} have a marked effect on them.

4. A New Roughness: Conjectures and a Picture

Formally, one might try to create an area reservoir sensitive to low tensions by assuming the bending rigidity to depend on the wavelength of an undulation mode. Evidently, for optically unresolved wavelengths the modulus would have to be at least two orders of magnitude smaller than the measured one. A bending rigidity that decreases with increasing wavelength is, however, difficult to imagine on physical grounds. New roughness can also arise from intrinsic or extrinsic defects in the membrane. The two simplest defects destroying planarity are hats and saddles; they have been considered in some detail in a previous publication [24]. Hats were shown to absorb plenty of area but at the same time to increase the flexibility, *i.e.* inverse rigidity, of the membrane in proportion to the absorbed area. On the basis of the measured rigidities hats were ruled out as the origin of more than a few percent of extra area. However, saddles were found to be capable of warping the membrane if their cooperativity is strong enough. They will affect the bending stiffness, making it larger or smaller, but need not destroy it.

A local saddle in an essentially flat membrane must be surrounded by two highs and two lows as sketched in Fig. 6. The complete saddle structure may be viewed as an excited state of the membrane if it represents a minimum of the bending elastic energy. Such a minimum can be brought about only be terms of higher than quadratic order in the curvatures. An estimate of the bending energy was based on the simplifying ansatz

$$g = \tfrac{1}{2}\kappa(c_1 + c_2)^2 + \bar{\kappa}c_1c_2 + \bar{\bar{\kappa}}(c_1c_2)^2$$

for the elastic energy per unit area of bilayer, where c_1 and c_2 are the principal curvatures. The first term, containing the bending rigidity, is the regular bending energy which will be large in the highs and lows.

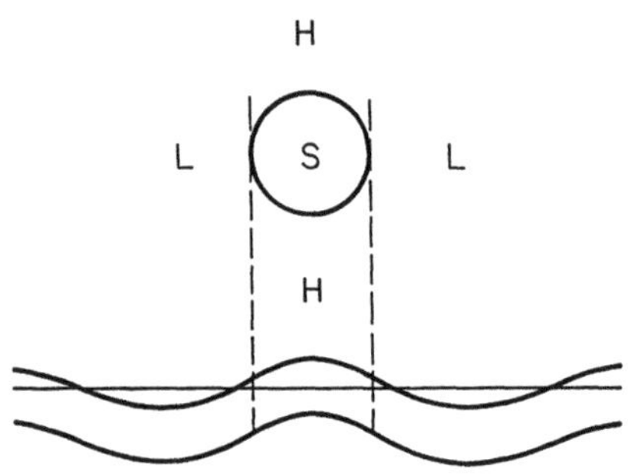

6. Local saddle with accompanying highs and lows, seen from above the membrane and in cross section through the lows.

The modulus of Gaussian curvature, $\bar{\kappa}$, may be positive or negative, but the second term does not affect the total energy of the saddle structure as the integral of Gaussian curvature over the membrane depends only on the boundary conditions, thus being the same before and after the deformation (Gauss-Bonnet theorem). The last term is quartic in the curvatures; with $\bar{\bar{\kappa}} < 0$ it lowers the bending energy of the saddles and, presumably to a lesser extent, of the highs and lows. Clearly, the fourth order term contributes to the total energy significantly only if the local curvatures are very large, but a natural lower limit for the radii of curvature is the monolayer thickness. The size of the saddle proper should be even smaller because the stronger the saddle curvature the smaller is the surface area over which it can be maintained.

The modulus $\bar{\bar{\kappa}}$ was calculated to be

$$\bar{\bar{\kappa}} \leq -\tfrac{2}{3}\gamma h^4$$

in terms of a particular microscopic model stemming from a theory of stretching elasticity proposed by Israelachvili *et al.* [32]. Here $\gamma \approx 50$ dyn cm^{-1} is the interfacial tension between hydrocarbon and water and $h = 2$ nm is the thickness of each of the two constituent monolayers. The desired negative sign results from assuming a stress profile of the bilayer like that shown in Fig. 7 in conjunction with an increase of the bilayer thickness under saddle curvature (= negative Gaussian curvature) to be expected with constant bulk density of the lipid. Crude estimates using $\kappa = 1 \cdot 10^{-12}$ erg then suggest for the total bending energy of the saddle structure

$$E_{\text{tot}} = \left(-1.3 \cdot 10^{-12}\text{erg} + 4 \cdot 10^{-12}\text{erg}\right) A_s/4\text{nm}^2.$$

The first term is the fourth-order bending energy of the saddle proper of membrane area A_s, while the second term stands for the regular bending energy of the highs and lows.

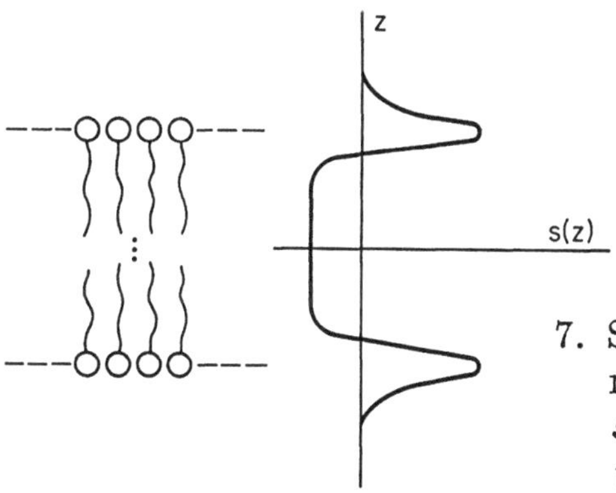

7. Schematic stress profile of electrically neutral biological model membranes. $s(z)$ represents the bulk normal tension (dyn cm^{-2}).

	S		H		S		H	S
	L		S		L		S	L
plane	S		H		S		H	S
	L		S		L		S	L
	(L)	S		(L)		S		(L)
large		(H)				(H)		
saddle	(L)	S		(L)		S		(L)
		(H)				(H)		
		(H)			(H)			
furrow or ridge	L	S		L	S	L		
		(H)			(H)			

8. Highly ordered arrays of saddle structures and the shapes they impose on the membrane in the large. The highs and lows in parentheses are thought to be weakly developed.

Two local saddles can share a high or a low, thus reducing the elastic energy of the combination. Accordingly, there will be a cooperativity of the saddles. Some highly ordered arrays that may result from it and the associated membrane shapes in the large are indicated in Fig. 8. Of particular interest in the present context would be a preference for large saddles and ridges as they might warp the membrane. A large saddle should require some defects in the lattice of small saddles. On the other hand, a regular lattice is acceptable if a few ridges join to form a thin tube.

Cooperative saddles might produce a roughness either on the scale of a few 100 nm or, perhaps, in the form of tubular protuberances which could both be sensitive to tensions as low as 10^{-5} dyn cm^{-1}. Increasing tension would first flatten the roughness but should finally annihilate the local saddles themselves. A simple estimate suggests that annihilation may happen in the vicinity of 10^{-2} dyn cm for the saddles structures of the type envisaged above. It could take a long time for new saddles to appear, even at zero tension, because of their (unknown) activation energy and their cooperativity. This might explain why electrically neutral biological model membranes behave "normally" [14,15], perhaps even display spontaneous adhesion [33], after being stretched by mechanical manipulation or an electric field.

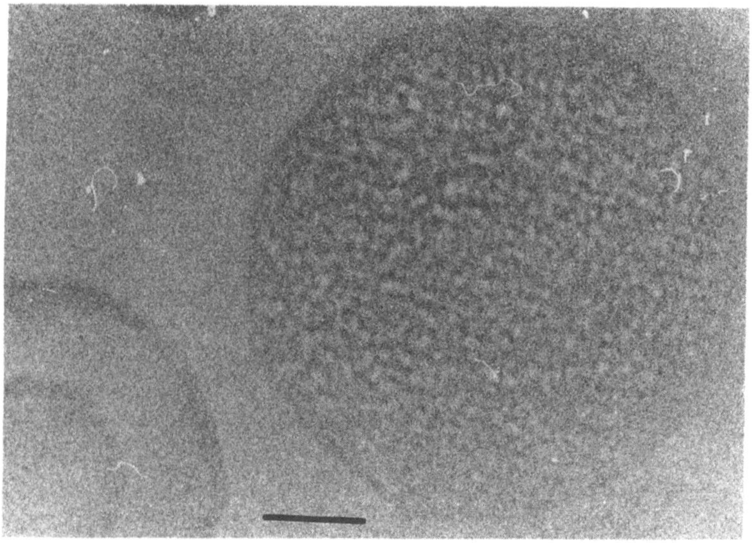

9. Micrograph of frozen egg lecithin vesicles obtained by transmission electron microscopy. The graininess of the vesicle shown fully may originate from both sides as the depth of focus comprises the sample thickness. Note a weak roughness of the membrane along the vesicle contour. The membrane of the vesicle in the upper right corner is smooth, as holds for the vast majority of vesicles. The bar represents 21.5 nm.

Since the beginning of 1988 we have been searching for the postulated roughness, mostly doing transmission electron microscopy on rapidly frozen thin samples of very dilute egg lecithin/water systems. Of a very large number of vesicles all were smooth except three. They exhibited not a roughness but a graininess which would be compatible with an array of densely packed saddles. One of them is shown in Fig. 9. The technical details are to be published elsewhere [25]. The light and dark dots on the membrane may be the lows and highs accompanying the saddles. We suspect that the local saddles of most of the vesicles were destroyed during sample preparation or rapid freezing, both of which may generate large lateral tensions. It was, in fact, this pattern which prompted one of us to develop the model of cooperative local saddles [24].

5. Conclusion

The contradictions of induced adhesion are not the only evidence for anomalous material properties of electrically neutral biological model membranes. First, it has been reported that the tubular vesicles of various lecithins, used in measurements of the bilayer bending rigidity, displayed intermittent wiggles and knees that appeared too strong to be explained in terms of regular bending elasticity [34]. Second, in all studies of the bending rigidity analyzing fluctuations the spread of the values obtained for the different vesicles or membranes seemed to be distinctly larger than the statistical error for the individual specimen. The scatter was particularly pronounced in the case of DGDG [35] where it may explain an enormous variation of the unbinding temperature among samples prepared from the same materials [23]. Third, Servuss has recently reported that the swelling of egg lecithin sometimes proceeds via the formation of close-packed parallel submicroscopic tubes [36]. Finally, there is evidence for a dispersed phase of egg lecithin in plenty of water at temperatures below 15° C which might consist of optically invisible suspended tubes [36].

One may wonder whether intrinsic defects other than saddles could warp the membrane. A possible candidate in this respect is the so-called lipidic particle which is assumed to be an inverse spherical micelle embedded in a bilayer [37]. Ridges and furrows with a row of lipidic particles in the middle have, in fact, been reported for some electrically neutral biological model membranes [37]. However, it is not clear why rows of such particles should bend a symmetric bilayer one way or the other and it has been suggested the rows actually are composed of local saddles [24].

In any event, biological model membranes promise to be much more "alive" than has been expected. Nature may have selected membranes with a high sensitivity to lateral tension and large contact angles of induced adhesion at the beginning of evolution when proteins were not as highly developed as they are today.

Acknowledgement

This work was supported in part by the Deutsche Forschungsgemeinschaft through SFB 312 and by the National Science Foundation under Grant No. PHY82-17853, supplemented by funds from the National Aeronautics and Space Administration. We also thank Evan Evans for continued discussions.

References

1. W. Harbich, H.J. Deuling and W. Helfrich, *J. Phys. France* **38**, 727 (1977).
2. J.F. Faucon, M.D. Mitov, P. Méléard, J. Bivas and P. Bothorel, *J. Phys. France* **50**, 2389 (1989); and references cited therein.
3. W. Helfrich, *Z. Naturforsch.* **30C**, 841 (1975).

4. F. Brochard, P.G. deGennes and P. Pfeuty, *J. Phys. France* **37**, 1099 (1976).
5. W. Helfrich and R.M. Servuss, *Il Nuovo Cimento* **3D**, 137 (1984).
6. S.T. Milner and S.A. Safran, *Phys. Rev.* **A36**, 4371 (1987).
7. W. Helfrich, *Z. Naturforsch* **33A**, 305 (1978).
8. W. Helfrich, *J. Phys. France* **48**, 285 (1987); and references cited therein.
9. L. Peliti and S. Leibler, *Phys. Rev. Lett.* **54**, 1960 (1985).
10. F. David, *Europhys. Lett.* **6**, 603 (1988).
11. P.G. deGennes and C. Taupin, *J. Phys. Chem.* **86**, 2294 (1982).
12. C.R. Safinya, D. Roux, G.S. Smith, S.K. Sinha, P. Dimon, N.A. Clark and A.M. Bellocq, *Phys. Rev. Lett.* **57**, 2718 (1986).
13. G. Porte, J. Marignan, P. Bassereau and R. May, *Europhys. Lett.* **7**, 713 (1988).
14. E. Evans, American Chemical Society Fall National Meeting, Miami, Sept. 10–15 (1989); and to be published.
15. M. Kummrow and W. Helfrich, to be published.
16. S.A. Safran, D. Roux, M.E. Cates and D. Andelman, *Phys. Rev. Lett.* **57**, 491 (1986).
17. D. Roux, M.E. Cates, M. Olsson, R.C. Ball, F. Nallet and A.M. Bellocq, ITP-UCSB preprint (August 1989).
18. R. Servuss, W. Harbich and W. Helfrich, unpublished, quoted in Ref. 8.
19. R.M. Servuss and W. Helfrich, *J. Phys. France* **50**, 809 (1989).
20. W. Harbich and W. Helfrich, *J. Phys. France*, submitted.
21. R.M. Servuss and W. Helfrich, in "Physics of Complex and Supermolecular Fluids," eds. S.A. Safran and N.A. Clark (Wiley, New York, 1987).
22. M. Mutz and W. Helfrich, to be published.
23. M. Mutz and W. Helfrich, *Phys. Rev. Lett.* **62**, 2881 (1989).
24. W. Helfrich, *Liq. Cryst.*, in press.
25. B. Klösgen and W. Helfrich, to be published.
26. W. Harbich and W. Helfrich, *Chem. Phys. Lip.* **36**, 39 (1984).
27. R. Kwok and E. Evans, *Biophys. J.* **35**, 637 (1981).
28. J. Marra and J. Israelachvili, *Biochemistry* **24**, 4608 (1985).

29. J. Mahanty and B.W. Ninham, *Dispersion Forces* (Academic Press, 1976).

30. R. Lipowsky and S. Leibler, *Phys. Rev. Lett.* **56**, 2541 (1986).

31. W. Helfrich, in "Phase Transitions in Soft Matter" (Nato Advanced Study Institute), ed. T. Riste (Plenum Press, 1989).

32. J.N. Israelachvili, D.J. Mitchell and B.W. Ninham, *J. Chem. Soc. Faraday Trans. II* **72**, 1525 (1976).

33. E. Evans and M. Metcalfe, *Biophys. J.* **46**, 423 (1984); and E. Evans, private communication.

34. G. Beblik, R.M. Servuss and W. Helfrich, *J. Phys. France* **46**, 1773 (1985).

35. M. Mutz and W. Helfrich, *J. Phys. France*, submitted.

36. R.M. Servuss, *Chem. Phys. Lipids* **50**, 87 (1989).

37. W. Helfrich and W. Harbich, in "Physics of Amphiphilic Layers," eds. J. Meunier, D. Langevin and N. Boccara, *Springer Proc. in Physics* **21** (1987); and W. Harbich and W. Helfrich to be published.

38. A.J. Verkleij, *Biochim. Biophys. Acta* **779**, 43 (1984); (review).

39. See, *e.g.*, R.P. Rand, *Ann. Rev. Biophys. Bioeng.* **10**, 277 (1981).

The Sponge State: A Striking Isotropic Liquid Phase

D. Roux [1,2] *and M.E. Cates* [1,3]

[1]Institute for Theoretical Physics, University of California, Santa Barbara, CA 93106, USA
[2]Centre de Recherche Paul Pascal, Domaine Universitaire, F-33405 Talence Cedex, France
[3]Cavendish Laboratory, Madingley Road, Cambridge, CB3 0HE, UK

Abstract. Surfactants in solution may associate into two-dimensional aggregates (membranes) which self-organize either in liquid crystalline phases or, more surprisingly, in a liquid isotropic phase made of connected bilayers. A review of what is known about this phase is presented. Structural and thermodynamical models of the phase are described together with experimental results.

1. Introduction

Due to their amphiphilic property, surfactant molecules in solution form aggregates which lead to a large variety of shapes and phases. Roughly speaking, one may distinguish between spherical (micelles), cylindrical (rod-like) and lamellar (disk-like) shapes. These supramolecular objects can organize themselves on a larger scale, either with long range ordering (liquid-crystal phases) or only short range correlation (liquid isotropic phases). Examples of liquid crystalline phases can be found involving either finite size micelles packed on a three-dimensional cubic lattice, infinite cylinders packed on a two-dimensional hexagonal lattice, or even infinite lamellae packed on a one-dimensional lattice (smectic A lyotropic phase).

These different liquid crystalline structures were recognized quite early [1] except for some of the cubic phases which can be quite complicated structures [2]. At first, the identification of a given shape in a liquid crystal phase is generally not too difficult, but to get the aggregate shape in liquid phases turns out to be a more difficult task. Only recently have non-spherical micelles been identified in isotropic liquid phases. Experimental data clearly demonstrate that cylindrical shapes (rods) occur. First thought of as rigid [3], these rods are now recognized as being semi-flexible [4] and this can give rise to very interesting polymer-like behavior [5,6]. The micellar to rod transformation is not a phase transition but rather a continuous smooth process that occurs upon changing the surfactant (or sometimes salt) concentration. The absence of a sharp transition is due to the one-dimensional nature of a rod and is related to the well-known result that there is no phase transition in one dimension [7]. Concerning the existence of two-dimensional objects in the liquid state (disks or lamellae) the situation is less clear. Although the existence of long range oriented phases (nematics) of disk-like micelles has been shown [8], isotropic liquid phases of surfactant packed in two-dimensional objects have been discovered only recently. Due to the very simple argument that the length of the edge of a disk grows linearly with its size, it is easy to show that phases of isolated disks are highly unstable (except in strongly interacting phases such as the nematic phase). In fact, if the surfactant molecules like to pack in a two-dimensional object it can be shown that one expects infinite films or membranes rather than

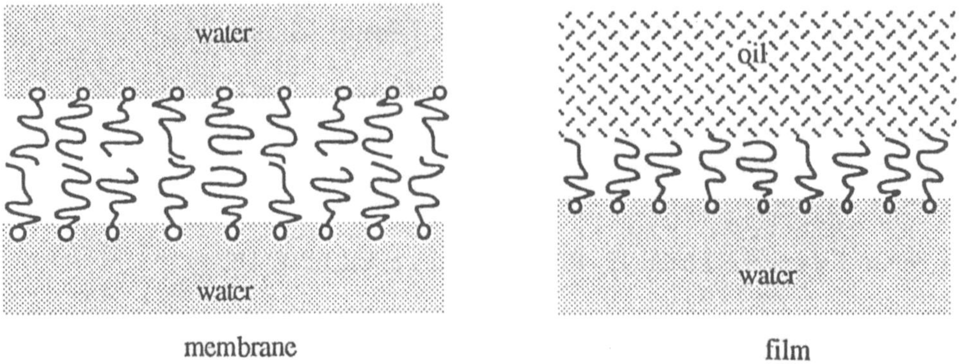

Figure 1. Schematic representation of surfactant two-dimensional organization: membranes are double layers of surfactants separating two identical media (water for example), and films are monolayers of surfactant separating two different media (oil and water).

finite disks [9]. In what follows we will be mainly concerned with two-dimensional aggregates (membranes or films) and we will see that a large variety of phases including isotropic liquids can be observed.

We will define a membrane as being constituted of a bilayer separating two identical media, in contrast to a surfactant film which is made of a monolayer separating two distinct media (see Fig.1). When the characteristic distance of the structure we are considering is large compared to the membrane (or film) thickness, the thermodynamics of this phase will be well described by the statistical physics of surfaces (2-D objects) embedded in a three-dimensional space. Arranging flexible surfaces with no edges in space is not easy. Two trivial solutions may be found: lamellar packing which can be described as a stack of infinite lamellae (lyotropic smectic A) or closed vesicles (Fig. 2a and b). The main difference between these two shapes is that in the first case the film is flat (average curvature $\langle H \rangle = 0$) while in the other case the film is curved ($\langle H \rangle \neq 0$). Another, less obvious, solution can be found: namely a phase of random surfaces that we will describe later (Fig.2c).

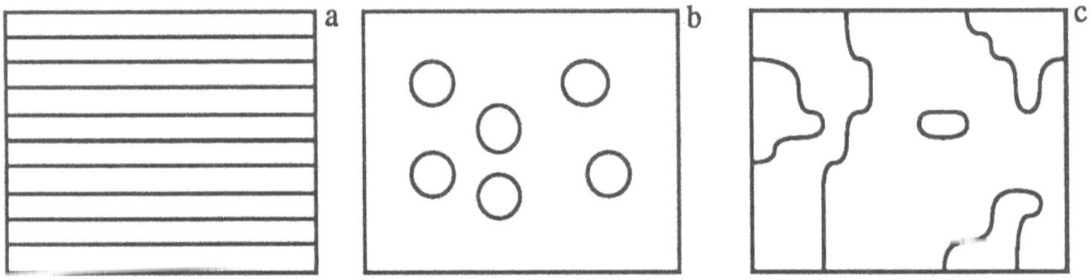

Figure 2. Two-dimensional section of long range organization for surfaces embedded in 3-d space. 2a represents a lamellar phase in which the membranes are packed in a one dimensionally long-range ordered phase. 2b is a liquid phase of spherical vesicles, and 2c a liquid phase of connected random bilayers.

2. Elasticity of membranes

In order to understand the statistical physics of these phases we must introduce the notion of elasticity of the film. Within the harmonic approximation the elastic energy of the surface can be described as [10]

$$E = \int \left[\mu + \kappa\left(H - H_0\right)^2 + \bar{\kappa}\, K\right] dS, \tag{1}$$

where μ is the chemical potential of the surface, H is the curvature ($H = 1/R_1 + 1/R_2$ where R_1 and R_2 are the two principal radii of curvature of the surface), and $K = 1/(R_1R_2)$ is the gaussian curvature. H_0, the spontaneous radius of curvature, is zero by symmetry for bilayers (membranes) but can be different from zero for films. The parameter κ is the mean bending constant and $\bar{\kappa}$ is the gaussian bending constant.

It can be shown (Gauss-Bonnet theorem) that

$$\int K\, dS = 4\pi\left(n_c - n_h\right), \tag{2}$$

where n_c is the number of disjoint pieces or components and n_h is the number of handles of the surface. Consequently, the gaussian curvature is a function only of the topology. The gaussian bending constant is then a chemical potential for the genus of the surface.

The rigidities κ and $\bar{\kappa}$ are both expressed in units of energy and this allows us to define two separate regimes concerning the statistical physics of membranes. In order to understand the effect of thermal fluctuations, one has to compare κ to k_BT (k_B is the Boltzmann constant and T the temperature): when $\kappa >> k_BT$, the membrane will be basically rigid and thermal fluctuations will have little effect on the shape of the membrane; on the other hand, when $\kappa \sim k_BT$ (flexible surfaces) thermal fluctuations will become important. For example, one effect of the thermal fluctuations is to renormalize the elastic constant κ and $\bar{\kappa}$. Within a perturbation theory it can be shown that [11-13]

$$\kappa = \kappa_0 - \frac{\alpha}{4\pi}\, k_BT \ln\left(\frac{q_{max}}{q_{min}}\right), \tag{3a}$$

$$\bar{\kappa} = \bar{\kappa}_0 + \frac{\beta}{6\pi}\, k_BT \ln\left(\frac{q_{max}}{q_{min}}\right), \tag{3b}$$

where α and β are numerical constants that can be calculated: according to [13], $\alpha = 3$ and $\beta = 5$. In (3), q_{max} and q_{min} are in-plane cutoffs corresponding to respectively the inverse of a molecular length (smallest allowed length scale of fluctuation) and the inverse of the in-plane correlation length of the fluctuations which, for a free membrane, corresponds to the transverse size of the membrane. This renormalization effect may be understood as due to the shape-dependence of the entropy of ripples (which is why the second term is proportional to k_BT) and the logarithmic dependence on q is a well known characteristic of fluctuations in 2-D systems. (This weak dependence on scale means that there is little effect, at length scales of interest, for bare rigidities large compared to k_BT.) The renormalization of the elastic constants has some important consequences; for example, the fact that the elastic constant decreases with the in-plane correlation length means that over a certain distance the membrane will lose its orientational order. From (3a) we are able to estimate this distance, called the persistence length of the membrane [14], as (crudely) the length at which the effective bending constant goes to zero. We get

$$\xi_k = a \exp\left(4\pi\, \kappa_0 / \alpha\, k_B T\right), \qquad (4)$$

with a being a molecular length. A membrane is consequently expected to be crumpled for lengths larger than ξ_k.

Thermal fluctuations are also responsible for an effect on lengths smaller than the persistence length. For a system of stacked membranes (lamellar phase), Helfrich has shown [15] that a long range repulsive interaction comes from the restriction of the undulations of the membrane (compared to a free membrane) resulting from steric exclusion by neighbours in the stack. In a regime where the distance d between membranes is smaller than the persistence length, the entropy lost per unit area associated with this restriction is

$$\frac{F}{A} = \frac{3\pi^2}{128} \frac{k_B T}{\kappa} \frac{k_B T}{d^2}, \qquad (5)$$

where the prefactor depends somewhat on the approximations made [15]. Experimental evidence for such an interaction has been recently demonstrated in systems of flexible membranes in the lamellar state [16-19]. The peculiarity of the systems studied is the large range of distances over which the lamellar phase is stable. Because of the long range repulsive interaction, these systems can be swollen by a large factor on adding solvent, and the repeat distance d of the smectic order can vary from a molecular length (50 Å) to very large distances (1000-10000Å) [20-22], allowing a quantitative check of both the d dependence and the amplitude in (5).

While it exists for a large range of concentrations (repeat distances) a lamellar phase stabilized by undulation interactions must eventually melt on dilution into an isotropic liquid phase. It may melt into a micellar phase through a strongly first order phase transition, but in the case of very stable bilayers, another possibility occurs. In fact, it has been proposed [23] and observed experimentally [24,25,26] that a phase of connected bilayers without long range positional or orientational order exists. Due to its particular structure, this phase is named the sponge phase. It is the structure and the thermodynamical properties of this phase that we would like to review in this article.

3. Structure and symmetry of the sponge phase

The membrane (bilayer) structure of the sponge phase has the following properties:

- the structure is made of self-avoiding connected bilayer, with a spatial distribution that is uniform at large distances;
- there is neither long-range positional nor orientational ordering of the membrane;
- the free energy cost of edges or seams in the bilayer is considered as prohibitive and therefore such defects are taken to be absent (see Fig.3).

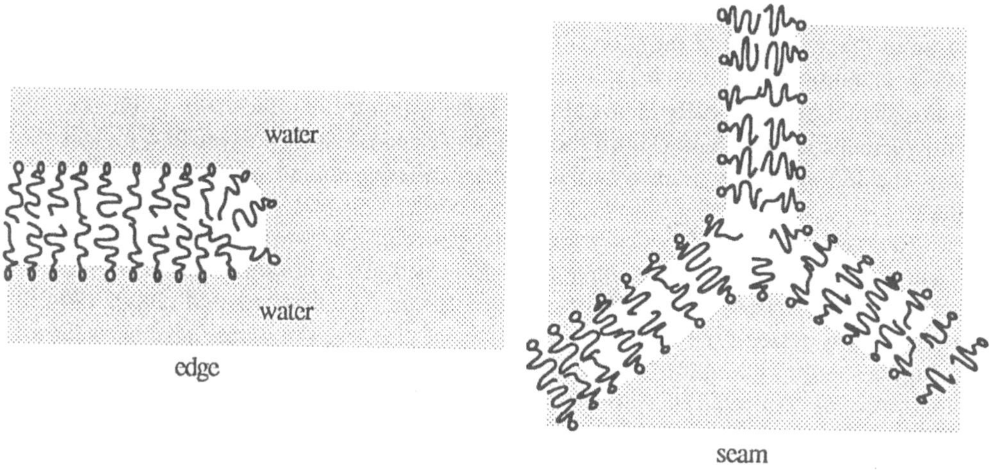

Figure 3. Typical defects that are not allowed in a sponge phase. The cost in bending energy of such defects is assumed to be prohibitive, because the surfactant film is bent on a molecular length scale (instead of a much larger one of order the persistence length).

It is possible to use a lattice model to generate such a geometrical arrangement of the surface [23]. The model corresponds to a slight modification of models used to describe bicontinuous microemulsion phases [27]. Assuming a random distribution of spins (up or down) on a three-dimensional lattice, one can define the interface between up and down domains. The membrane is assumed to lie on this interface without allowing self-crossings (no seams). We can then remove the spins and keep the membrane; as shown in Fig. 4, the resulting surface divides space into two bicontinuous regions of identical material (solvent) arbitrarily designated as I (Inside) and O (Outside). This labelling is unique up to a global interchange between I and O. The hamiltonian must be invariant under this interchange

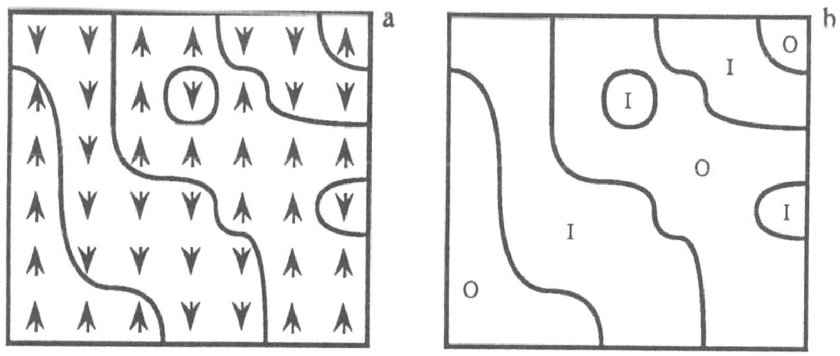

Figure 4. Lattice construction of a random phase of bilayers (sponge phase). The lines represent the interface between domains of a given spin orientation. In the model they are assumed to be surfactant bilayers separating two identical media (water for example) labelled I (In) and O (Out).

because of the symmetry of the bilayer about its mid-plane; this is true for the harmonic elasticity hamiltonian (eq. 1 with $H_0 = 0$) but also for more general forms. Thus the I and O domains should be statistically identical unless the symmetry is *spontaneously* broken. [In the microemulsion case I and O correspond to water and oil; a monolayer film divides the space. The I/O symmetry is then broken by the different nature of the two continuous media.]

There is experimental evidence for the existence of a symmetric sponge phase. Both oil- and water-rich sponge phases have been described. They are found in the vicinity of swollen lamellar phases stabilized by undulation forces. The first reported phases [24-26] were quaternary systems but even binary (surfactant plus water) systems have been shown to exhibit a sponge phase [28]. The experimental facts leading to a geometrical description of a structure of connected bilayers are based on two measurement techniques: transport properties using conductivity and structural determination using neutron scattering.

The most spectacular conductivity measurements are found in an oil-rich sponge phase [26]. This particular phase is stable in a surfactant concentration range between 20 to a few percent. The solvent is a mixture of dodecane and pentanol and is an insulator. The membrane is an inverted bilayer, slightly swollen with water, which can be considered as a conductor (the water is filled with charges coming from the surfactant counterions). Measurements of the reduced conductivity (compared to the expected conductivity for the same amount of pure water + charges) shows a high conductivity (around 0.65) all along the dilution path, even for the most dilute samples. A comparison with the conductivity of a nearby droplet phase (inverted micelles) for the same surfactant concentration shows order of magnitude differences in the conductivity (of the order 10^{-1} - 10^{-4} for the droplet phase). This indicates that the bilayers are connected and that continuous water paths through the membranes can be found. The value of .65 instead of 1 (expected for example in perfect lamellar phase along the direction parallel to the layers) is easily explained in terms of the tortuosity of the membrane path.

In a water-rich sponge the conductor is the solvent, and the insulator is the membrane; in this case the reduced conductivity is found to be 2/3 of the conductivity of the pure solvent [24,26]. Again the ratio of 2/3 can be interpreted as related to the tortuous path through a dielectric medium filled with randomly distributed insulating surface elements.

Other evidence for the sponge structure has been obtained by neutron scattering experiments. Typical data of a series of samples along a dilution line in absolute units are shown in Fig. 5. The scattering intensity can be divided into two regimes: the large q regime where all the data fall on a universal curve (which is the form factor of a flat bilayer of the same thickness as that measured in the lamellar phase) and the small q regime where correlations between bilayers lead to a broad peak. The large q regime crosses over a $1/q^2$ behavior for $q<2\pi/\delta$ (δ is the bilayer thickness), which is expected for the scattering of a flat bilayer with random orientation, and a $1/q^4$ behavior at larger q ($q>2\pi/\delta$) characteristic of the scattering from a thin interface. The position of the broad peak due to bilayer-bilayer correlations also gives structural information. In a lamellar phase the quasi-Bragg peak due to the long range order of the bilayers can be exactly located in q-space (neglecting logarithm corrections due to the undulations) as $q=2\pi/d$ with $d= \delta/\phi$ (ϕ is the volume fraction of surfactant). For a random distribution of connected bilayers, one expects $d = \gamma\delta/\phi$ with γ a number larger than 1 (the lattice model leads to $\gamma = 1.5$). The peak position for the sponge phase indeed shows a $1/\phi$ behavior with $\gamma = 1.4 - 1.6$ depending on the system.

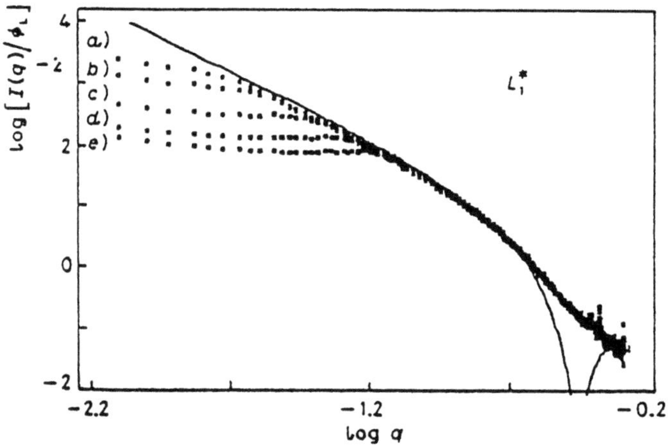

Figure 5. Neutron data of a series of samples in a sponge phase. The scattering intensity has been recorded in absolute units and the reduced intensity (I/ϕ) is plotted against the q vector.

4. Stability and thermodynamical properties

The stability and thermodynamical properties of the sponge phase can be obtained by using microscopic models or more general descriptions (Landau theory for example). We will first examine the properties of the model previously described, and then we will make some more general statements.

4.1 Lattice model

The geometrical model described in the previous section has been used to calculate the free energy of the sponge phase in a random mixing approximation [23,29]. The free energy has two contributions: the entropy of the membrane (arising from its random spatial arrangement) and the elastic free energy. Since the I/O symmetry of the hamiltonian may break spontaneously, the ratio of spin up to spin down regions (I/O ratio) need not be unity, in general. A realistic model should allow the system to choose for itself a ratio between I and O. This can be taken into account in the lattice model by introducing a parameter ψ for the amount of (say) I, in which case $(1-\psi)$ is the amount of O. The free energy can be written as a function of variables ψ and ϕ (the surfactant volume fraction) and then it can be minimized with respect to ψ. The mean curvature can be calculated by examining only 4-spin configurations, at the level of a random mixing approximation [23,27]. This leads to an elastic energy density

$$F_{ben} = \lambda \frac{\kappa(\xi)\,\psi\,(1-\psi)}{\xi^3}, \qquad (6)$$

where ξ is the cell size, related to the amount of surface (surfactant concentration ϕ) by

$$\xi = \frac{6\,\psi\,(1-\psi)}{\phi}\,\delta\,; \qquad (7)$$

λ is a geometrical parameter defining the amount of curvature per corner for the lattice model. If we assume that curvature energy of an isolated cube should, realistically, reduce to that of a sphere, then $\lambda = 8\pi$. Above, δ is the bilayer thickness and $\kappa(\xi)$ is the bending constant,which is scale-dependent as expected from (3a), which may be rewritten

$$\kappa(\xi) = \kappa_0 - \frac{\alpha\, k_B T}{4\pi} \ln(\xi/\delta)\,. \tag{8}$$

The question of the gaussian elastic energy is more difficult to address. Since the phase transition between the lamellar phase and the sponge phase involves a change in topology one has, in principle, to consider a contribution of the gaussian term in (1) [25]. The first proposed models were studied mainly for $\bar{\kappa} = 0$ and consequently the gaussian contribution to the elastic energy was not included [23,27]. The evaluation of the gaussian curvature in the lattice model requires a counting of 8 spin contributions [30]. There is some ambiguity concerning the topological assignment of edges where two I and two O domains meet diagonally; with the simplest reasonable assumptions, we found [31]

$$F_{gau} = \frac{4\pi\, \bar{\kappa}\, \psi(1-\psi)}{\xi^3} \left[1 - 16\,\psi\,(1-\psi)\left(\psi^2 + (1-\psi)^2\right)\right]. \tag{9}$$

The gaussian curvature is negative for $\psi = 1/2$, indicating that in the symmetrical sponge phase the number of handles is larger than the number of disconnected components.

The entropic contribution to F is dominated by the entropy of mixing of the I and O domains and can be easily estimated in a random mixing approximation:

$$-\frac{TS}{V} = \frac{k_B T}{\xi^3}\left[\psi \ln(\psi) + (1-\psi)\ln(1-\psi)\right]. \tag{10}$$

The total free energy density F is obtained by summing (6), (9) and (10). Note that the entropy of the film is evaluated in two parts: for length scales larger that the cell size it is given by the mixing entropy of I and O domains (10) and for length scales smaller than the cell size it contributes to the reduction of the bending constant (8).

The evaluation of the stability of the sponge phase requires a comparison with the lamellar phase. The lamellar free energy is simply described by a system of interacting membranes assuming the undulation interactions to dominate, according to (5) :

$$F_{lam} = \mu \frac{(k_B T)^2}{\kappa\, d^2}\,. \tag{11}$$

d is the repeat distance in the lamellar phase ($d = \delta/\phi$) and μ is a numerical coefficient [23]. The lamellar phase free energy is evaluated separately and the two branches of free energy are compared [23]. A better treatment requires a continuous description of the lamellar to sponge phase transition. Golubovic and Lubensky have proposed such a treatment and have included the steric repulsion in the sponge free energy [29]. However our simpler approach, which in effect assumes a strongly first order phase transition to lamellae, captures the basic features. In what follows we will not include steric interactions in the sponge phase.

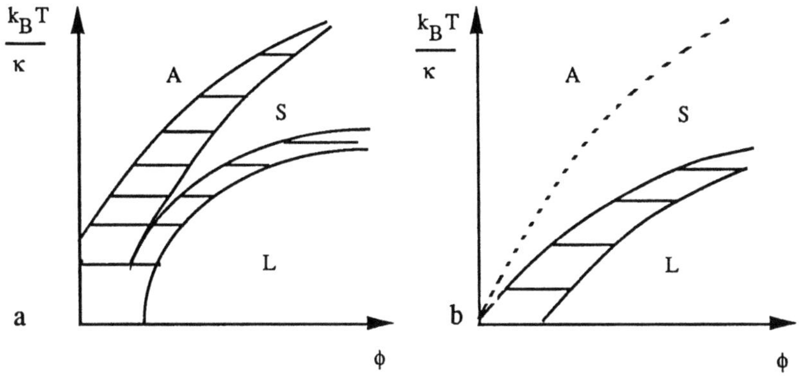

Figure 6. Typical phase diagrams calculated with the free energy given by (6), (9), (10). L is a lamellar phase, S the symmetric sponge phase and A an asymmetric sponge or (when very dilute) a phase of small vesicles. The phase transition between the symmetrical sponge and the asymmetrical sponge can be either first order (6a) or second order (dashed line 6b).

The thermodynamics is obtained after minimisation of F with respect to ψ. Typical phase diagrams (obtained for $\bar{\kappa} = 0$) are shown in Fig. 6. Depending on the reduced temperature and the volume fraction of surfactant, three phases are present: a lamellar phase (L), a symmetrical sponge (S) in which the parameter ψ is equal to 1/2 (same amount of I and O) and an asymmetric sponge phase (A) where ψ is different from 1/2. (This phase reduces to a dilute vesicle or droplet state at very low volume fractions.) One important property of such a phase diagram is to exhibit a phase transition between a symmetrical sponge and an asymmetrical phase of bilayers. This phase transition can be either first order (Fig.6a) or second order (Fig.6b) [29] and can be identified with the tense-relaxed phase transition described by Huse and Leibler [32]. When second order, this phase transition is expected to be Ising-like [32,33]. The effect of the gaussian curvature can be also investigated, and neglecting the renormalization of the gaussian bending constant the phase diagram at a given temperature (T/κ) is shown in Fig.7. As illustrated in Fig.7, tuning $\bar{\kappa}$ at constant temperature is enough to produce the sponge - lamellar phase transition.

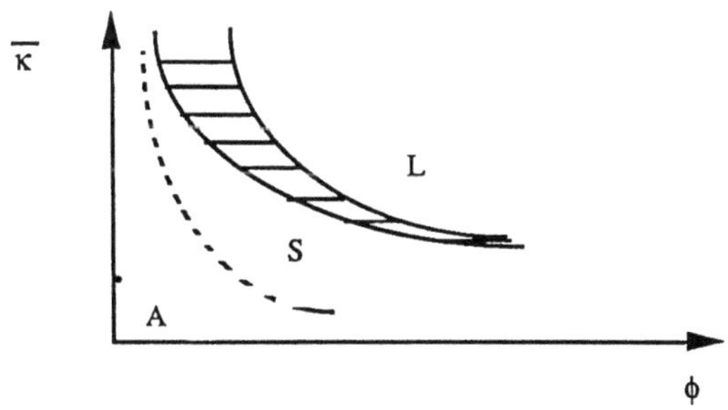

Figure 7. Phase diagram of the symmetric sponge (S),lamellar (L) and asymmetric (A) phases. The reduced temperature is kept fixed (k_BT/κ) and the gaussian bending constant is varied.

The free energy of the symmetrical sponge according to this model is simply

$$F = \phi^3 \left[A + B \ln(\phi)\right] \quad (12)$$

with $A = \frac{8\,k_B T}{27\,\delta^3}\left[\ln(2) + \frac{\lambda}{4}\left(\frac{\alpha}{\pi}\ln\left(\frac{3}{2}\right) + \frac{\kappa_0}{k_B T}\right) - \frac{\pi\,\bar{\kappa}}{k_B T}\right]$ and: $B = \frac{8\,k_B T}{27\,\delta^3}\left[\frac{\lambda\,\alpha}{4\pi}\right]$.

As shown by Huse and Leibler [32] and Porte et al.[25], the ϕ^3 dependence of F is a very general property coming from the scale invariance of the bending energy of a surface. However, this invariance is broken by the entropy of the small ripples which give logarithm corrections. More generally, A and B are coefficients which are expected to be functions only of κ, $\bar{\kappa}$, and T. The ϕ^3 scaling argument holds only if "decorations" of the basic structure are negligible. By decorations, one means the presence of extra structures of size much smaller than the cell size ξ (such as small vesicles or micelles floating inside the sponge) which contribute to the free energy with a different ϕ-dependence. While there is no theoretical demonstration that decorations can be safely ignored, the experimental measurements agree well with an assumption that their effects are negligible.

4.2 Landau theory

A main feature of the previous model is that due to the special symmetry of the sponge phase (I/O) the free energy depends on two "order parameters". One is the volume fraction of membrane ϕ (or its departure from a critical value $\rho = \phi - \phi_c$) and the other one is the parameter ψ, or rather its deviation from 1/2: $\eta = \psi - 1/2$. Since the symmetric sponge phase (at least in the model described) exhibits an exact I/O symmetry, we must expect there to always be a phase transition between the symmetrical sponge and another liquid isotropic phase, existing at smaller concentrations, in which the symmetry is broken. The two order parameters are expected to be coupled, as in fact they are in the lattice model. This description is related to the Blume-Emery-Griffith model for the He^3-He^4 phase diagram [34]. At a general level the problem of the expected phase diagrams can be addressed by a Landau theory involving our two coupled order parameters: the density ρ and the Ising-like order parameter η [35,36]. Assuming only the lowest order coupling term permitted by I/O symmetry, we have

$$\Delta F = \mu\,\rho + \lambda\,\rho^2 + \alpha\,\rho^4 + \beta\,\eta^2 + \delta\,\eta^4 + \nu\,\rho\,\eta^2 \quad (13)$$

where μ, λ, α, β, δ and ν are coefficients that can be estimated from a specific microscopic theory. It can be shown [36] that under variation of these coefficients, 4 basic types of phase diagrams arise. Figure 8 exhibits the different phase diagrams obtained with this Landau free energy. As shown, the phase transition between the sponge and the dilute phase can be either first order or second order; in addition, a liquid gas critical point can exist either between two symmetric sponge phases or two broken symmetry phases. The phase diagrams obtained with more specific models [23, 29] belong to one or other of the basic types. The I/O symmetry-breaking phase transition is governed by the order parameter η, whereas the liquid gas critical point is associated with the density order parameter ρ. Since these are

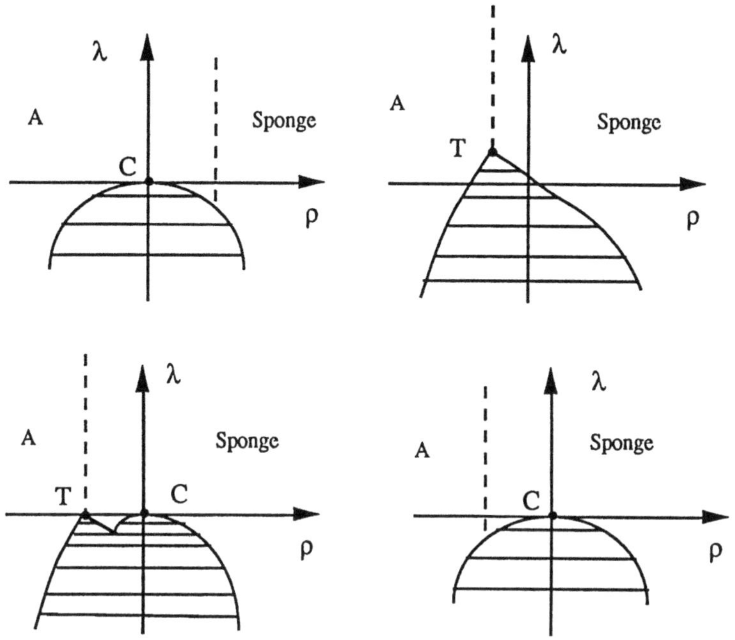

Figure 8. Typical phase diagrams obtained with the Landau free energy given by (13). The symmetric sponge to asymmetric sponge transition can be either first order or second order. T is a tricritical point and C a liquid-gas-type critical point.

coupled, we shall see below that ρ has very interesting behaviour near the I/O transition as well as the usual liquid gas behavior at its own critical point.

4.3 Correlation functions

Using Landau Ginsburg theory we can work out the expected correlation functions in the symmetric sponge phase, near the second order transition line to the asymmetric (broken symmetry) phase. Experimentally the only observable fluctuations are in the density, since there is no field able to couple to the I/O order parameter directly. However, as we shall see, there are some interesting effects of the η fluctuations on the density-density correlation functions [35,36]. In order to calculate the correlation functions we have to add to eq. 13 suitable gradient terms. We assume that we are in the symmetric sponge state, approaching the critical I/O line, but not too closely, in which case we may keep a gaussian treatment of the fluctuations. So long as $\lambda > 0$, we can write

$$H = \mu\,\rho + \lambda\,\rho^2 + \beta\,\eta^2 + \nu\,\rho\,\eta^2 + \gamma\,(\nabla\eta)^2 + \tau\,(\nabla\rho)^2 \qquad (14)$$

neglecting, for simplicity, a cross-term in the gradients, and also higher order gradient terms (as they arise) below. The I/O order parameter η is then found to have gaussian fluctuations of Ornstein-Zernicke type

$$\langle \eta(0)\eta(r) \rangle = \frac{e^{-r/\xi_\eta}}{8\pi\gamma r}, \tag{15}$$

where $\xi_\eta^2 = \gamma/(\beta - \mu\nu/2\lambda)$. The surfactant density-density correlations come from two contributions: the intrinsic fluctuations of ρ controlled by λ, and the fluctuations induced by the coupling to η in (14). We can show that [35,36]

$$g(r) = \langle \rho(0)\rho(r) \rangle = \frac{e^{-r/\xi_\rho}}{8\pi\tau r} + \frac{\nu^2}{2\lambda^2} \frac{e^{-2r/\xi_\eta}}{(8\pi\gamma r)^2}, \tag{16}$$

where $\xi_\rho = (\tau/\lambda)^{1/2}$ is the correlation length for the intrinsic fluctuations of ρ, and ξ_η is the correlation length for the fluctuations of η. The Fourier transform of g(r) gives a prediction for the q behavior for light scattering intensity measurements (small q) in the sponge phase. We get [35]

$$I_s(q) = A\left[\frac{B}{1+\xi_\rho^2 q^2} + \frac{\tan^{-1}(q\xi_\eta/2)}{q\xi_\eta/2}\right], \tag{17}$$

where A and B are functions of the coefficients of the Landau Ginsburg free energy. The first term of (17) is just a regular Ornstein-Zernike behavior expected for intrinsic density fluctuations of ρ alone; the second term comes indirectly from the coupling to fluctuations of η. Depending upon B, and the relative magnitude of the correlation lengths ξ_ρ and ξ_η (i.e. on the relative distance from the I/O critical line as opposed to the liquid-gas critical point), we may expect one or the other term to dominate the experimental behavior. Experimental evidence of this unusual scattering was found in a quaternary system [35]. In this particular case, ξ_η was found to be much larger than ξ_ρ and in fact the experimental data (static light scattering) could be fitted by expression (17) assuming $\xi_\rho = 0$ (Fig. 9).

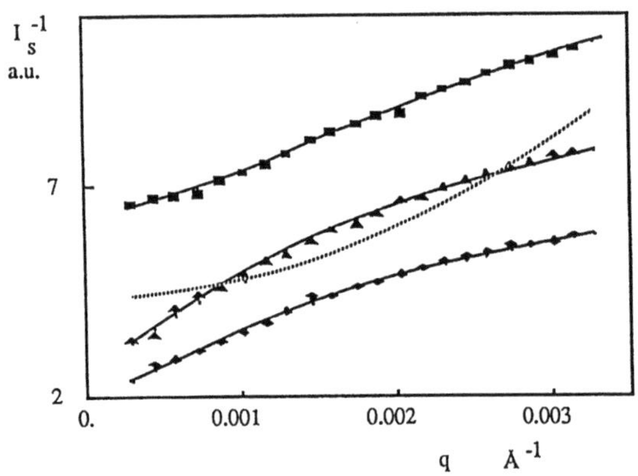

Figure 9. Plot of the inverse scattering intensity against q for a sponge phase at different concentrations. The dotted line is a fit using a normal Orstein-Zernike formula, whereas the full lines are fits using (17) (with $\xi_\rho = 0$).

From the experimental fit, the osmotic compressibility of the symmetric sponge phase can be obtained. We have seen that the free energy of the sponge is expected to scale as ϕ^3 with logarithm corrections (see eq. 12). Consequently the light scattering intensity at zero wave vector will scale as

$$I(q=0) \sim \phi \left(\frac{\partial \pi}{\partial \phi}\right)^{-1} \sim \frac{1}{\phi \ln\left(\frac{\phi}{\phi^*}\right)}, \tag{18}$$

where ϕ^* is approximately the concentration on the critical line for the symmetrical to asymmetrical phase transition (or the location of the spinodal line in the case of a first order phase transition). Figure 10 shows the experimental behavior for the light scattering intensity; it is clear from the plot that the theoretical prediction including the logarithmic correction is very well confirmed.

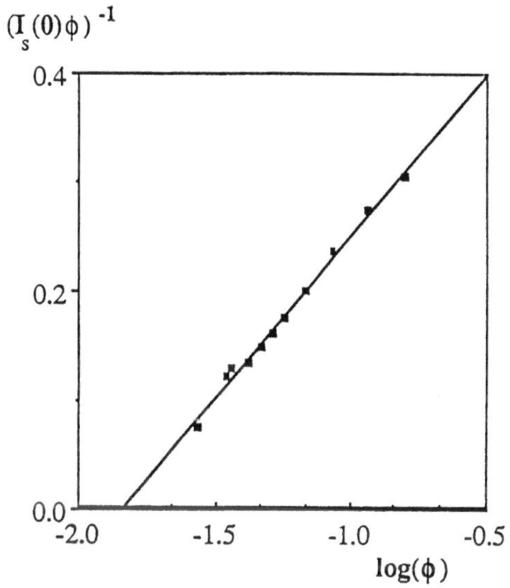

Figure 10. Plot of the inverse of the reduced scattering intensity at q=0 as a function of log(ϕ) for a sponge phase varying the volume fraction. The linear behavior is due to logarithm corrections of the compressibility (see eq. 18).

5. Conclusions

We have seen that, in certain cases, the existence of an isotropic liquid phase made of random connected bilayers is expected. This membrane divides space into two bicontinuous domains arbitrarily labelled In (I) and Out (O). In the case where the amounts of I and O are equal the two bicontinuous domains are statistically identical and the phase is symmetrical with respect to I and O. This symmetry allows us to define an order parameter, which can be for example the average curvature $\langle H \rangle$: $\langle H \rangle = 0$ in the symmetrical sponge and $\langle H \rangle \neq 0$ in an asymmetrical sponge. This special symmetry may be spontaneously broken, but even in the unbroken (symmetric) phase it leads to some novel effects. For example, the concentration fluctuations for the surfactant are not Orstein-Zernik-like but have a strikingly different q dependence (at least in a certain domain of the phase diagram) due to the coupling between the I/O order parameter and surfactant concentration.

The existence of this special symmetry is directly linked to the suppression of defects within the bilayer such as edges or seams [23]. For the symmetry to be well-defined we must assume the nonexistence of such defects; however, in real systems we may expect them to arise at some finite density. To effectively destroy the distinction between I and O regions (without which the symmetry is presumably meaningless) the defects must be quite large -- of the order of the sponge characteristic length ξ. A rough estimate of the energy cost of such defects is $\kappa\, \xi^2/\delta^2$ where δ is the membrane thickness. Since ξ can be many orders of magnitude larger than δ, we may expect this energy cost to be quite big compared to the thermal energy, and consequently to have an average distance between defects L that is much larger than ξ. However, the presence of a finite density of defects may have some important consequences for the continuous phase transition described previously. For example, instead of a true divergence of the thermodynamical properties at the transition as expected for an Ising-like description, we might observe a rounding off when the critical length of fluctuations is of the order L. Of course, this rounding may be unobservable if L is large enough.

Setting aside this effect (which is rather like a finite-size correction) we may expect to observe, in suitable systems, very interesting new behavior such as the existence of a critical line between two isotropic liquids, and the existence of a symmetrical tricritical point (even for binary systems). The data presented above are for a system in which the I/O transition is actually first order, but this is not expected on any fundamental grounds since (by I/O symmetry) there are no odd powers of η in the Landau expansion. [Note that the coupling to ρ does not lead to a first order transition, although it would do so in a system of finite shear modulus [37]]. Thus systems should exist in which the transition is second order over a wide temperature range; a candidate is the water-rich analogue of the quaternary one described above [26], and work on this system is in progress. In principle, since one is working in a state of fixed surfactant concentration (rather than fixed chemical potential) the Ising exponents should be Fisher-renormalized at the second order line [37]. The dynamical properties of the sponge phase, especially near the I/O symmetry breaking line, are also the subject of current theoretical and experimental study.

Acknowledgements

Most of the work and ideas presented here have been shared by many persons. We would like to specially thank S. T. Milner, D. Andelman, S. Safran, R. Ball, A.M. Bellocq, D. Gazeau, F. Nallet, U. Olsson, and T. Zemb. This work was funded in part by NSF Grant No. PHY82-17853 supplemented by funds from NASA at the University of California in Santa Barbara.

References

1 P. Ekwald in "Advances in Liquid Crystals", G. H. Brown,ed., (Academic Press, NY) 1975

2 J. Charvolin, J. Chem. Phys. 1, **80** (1983)

3 P. J. Missel, N. A. Mazer, G. B. Benedek and M. C. Carey, J. Phys. Chem. **87**, 1264 (1983)

4 G. Porte, J. Appell and Y. Poggi, J. Phys. Chem. **84**, 3105 (1980)
5 S. J. Candau, E. Hirsh and R. Zana, J. Coll. Interf. Sci., **105** 521 (1985)
6 M. E. Cates, J. Physique., **49**, 1593 (1988)
7 H.E. Stanley, "Phase Transitions and Critical Phenomena" (Oxford University Press Oxford 1971).
8 L. J. Yu and A. Saupe, Phys. Rev. Lett. **45**, 1000 (1980)
9 J. Israelashvili, D.G. Mitchell and B. Ninham J. Phys. Chem. **72**, 1525 (1976)
10 W. Helfrich, Z. Naturforsch., **28c**, 693 (1973)
11 W. Helfrich J. Physique , **46**, 1263, (1985)
12 L. Peliti and S. Leibler, Phys. Rev. Lett. **54**, 1690 (1986)
13 F. David in "Statistical mechanics of membranes and surfaces" D. Nelson, T. Piran and S. Weinberg eds. (World Scientific, Singapore 1989).
14 P. G. de Gennes and C. Taupin, J. Phys. Chem. **86**, 2294 (1982)
15 W. Helfrich, Z. Naturforsh, **33a**, 305 (1978)
16 C. R. Safinya, D. Roux, G. S. Smith, S. K. Sinha, P. Dimon, N.A. Clark and A. M. Bellocq, Phys. Rev. Lett.. **57**, 2718 (1986)
17 D. Roux and C. Safinya, J. Physique , **49**, 307 (1988)
18 F. Nallet, D. Roux and J. Prost, Phys. Rev. Lett. **62**, 276 (1989)
and J. Physique **50**, 3147 (1989)
19 C. R. Safinya, E. Sirota, D. Roux and G. Smith, Phys. Rev. Lett. **62**, 1134 (1989)
20 F. Larche, J. Appell, G. Porte, P. Bassereau and J. Marignan, Phys. Rev. Lett. **56**, 1700 (1986)
21 J. Appell, P. Bassereau, J. Marignan and G. Porte, to be published
22 P. Bassereau, J. Marignan and G. Porte, J. Physique , **48**, 673, (1987)
23 M. E. Cates, D. Roux, D. Andelman, S. Milner and S. Safran , Europhys. Lett. **5**, 733 (1988); erratum ibid **7**, 94; M. E. Cates, proceedings of NATO ASI on "Phase Transitions in Soft Condensed Matter" D. Sherrington and T. Riste, eds., (Plenum, NY , 1989)
24 G. Porte, J. Marignan, P. Bassereau and R. May, J. Physique , **49**, 511, (1988)
25 G. Porte J. Appell, P. Bassereau and J. Marignan, J. Physique , **50**, 1335 (1989)
26 D. Gazeau, A. M. Bellocq, D. Roux and T. Zemb Europhys. Lett. **9**, 447 (1989)
27 D. Andelman, M. E. Cates, D. Roux and S. Safran, J. Chem. Phys. **87**, 7229 (1987); see also Phys. Rev. Lett. **57**, 491 (1986).
28 R. Strey, R. Shomaker,. D. Roux, F. Nallet and Olsson, J. Chem. Soc. Far. Trans., to be published
29 L. Golubovic and T. Lubensky Europh. Lett. **10**, 513 (1989)
30 H. Kleinert J. Chem Phys., **86**, 3565 (1987)
31 D. Roux, M. E. Cates unpublished.
32 D. Huse and S. Leibler J. Physique., **49**, 605 (1988)
33 F. David, Europhys. Lett. **9**, 575 (1989)
34 M. Blume, V. J. Emery and R.B. Griffiths, Phys Rev. A **4**, 1071 (1971)
35 D. Roux, M. E. Cates, U. Olsson, R.G. Ball, F. Nallet and A.M. Bellocq, Europhys. Lett., **11**, 229 (1980)
36 D. Roux, C. Coulon and M. E. Cates, to be published
37 D. Bergman and B. Halperin, Phys. Rev. B **13**, 2145 (1976)

Pattern Formation Processes in Binary Mixtures with Surfactants

T. Kawakatsu and K. Kawasaki

Department of Physics, Kyushu University 33, Fukuoka 812, Japan

Abstract. A dynamical model is proposed which is suitable for investigating the pattern formation processes in binary mixtures with surfactants. Typical examples of such systems are microemulsions. Bicontinuous and micellar pattern formation processes are demonstrated and the growth laws of the patterns are studied by computer experiments.

1. Introduction

A surfactant is a molecule which is composed of two distinct parts and has an amphiphilic nature. In an immiscible binary mixture, the surfactants align on the interfaces between two phases and lower the surface tension of the interfaces. If the surface tension becomes nearly zero, there emerges a thermodynamically stable mesophase with a characteristic length of the order of 100 Å. Such a phase is called a microemulsion [1]. The static properties of the microemulsion phase have been extensively investigated both experimentally and theoretically. On the other hand, the problems of the dynamical aspects of microemulsions are still open.

2. Models and Computer Experiments

We propose a model which is suitable for investigating, with use of computer experiments techniques, the dynamical processes of immiscible binary mixtures with surfactants, such as the formation of micelles (globular domains), bicontinuous structures (percolated random domains) and lamellar structures (layered structures). In our model we represent the binary mixture by a continuous scalar field, while we treat the surfactant molecules as discrete particles with directors interacting with the field and with each other. We combine the cell dynamics method [2] and the molecular dynamics method [3] to integrate the equations of motion of the system. Using such a hybrid model we computed the time evolution of the system from the initial configurations where the oil, water and surfactants are randomly mixed and we can obtain the micellar phase and the bicontinuous phase similar to those in the microemulsion phase [4] (see Fig.1). We computed the characteristic length of the pattern and found that the domain growth processes in our emulsion systems are different from those of the quenched alloy systems. It is characteristic of the emulsion systems that

Springer Proceedings in Physics, Vol. 52 **Dynamics and Patterns in Complex Fluids**
Editors: A. Onuki · K. Kawasaki

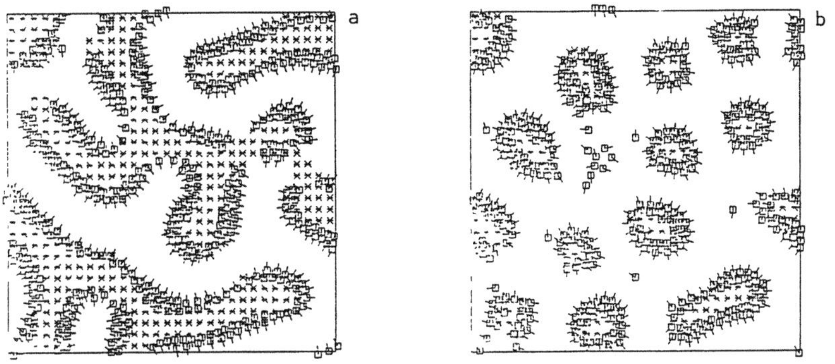

Fig.1 a) Bicontinuous and b) micellar patterns produced by the computer experiments are shown. The system is divided into 30×30 cells and 500 surfactants are immersed.

the surfactant molecules are adsorbed on the interfaces and that they prevent the interfaces from moving or shrinking.

Besides the hybrid model mentioned above, we can construct a continuum model in which the surfactants are also described by two continuous variables, one is the number density of the surfactants and the other is the vector field defined by the local average of the directors. This continuum model can be derived from the above-mentioned hybrid model by taking the continuum limit of the surfactant distribution. Assuming the strong segregation limit where the sharp interfaces are formed, we can show using the continuum model that the interfacial free energy has the usual form including bending elasticity obtained by Helfrich [5]. This continuum model can also describe the dynamics of the emulsion systems on a coarse-grained scale and, therefore, can be a basis of the studies of the macroscopic pattern formation processes.

3. Conclusion

In this paper we showed that our hybrid model is suitable for investigating dynamics of the emulsion system. We also sketched a continuum model which will be complementary to the hybrid model. These two models can provide the basis for investigating the dynamics of the emulsion system.

The authors would like to thank Dr. K. Sekimoto for valuable discussions.

References

1) P.G. De Gennes and C. Taupin: *J. Phys. Chem.*, **86** (1982) 2294.
2) Y. Oono and S. Puri: *Phys. Rev.*, **A38** (1988) 434.
3) M.P. Allen and D.J. Tildesley: *Computer Simulation of Liquids* (Oxford, 1989).
4) T. Kawakatsu and K. Kawasaki: to be published in *Proceedings of the Yamada Conference on Strongly Coupled Plasma Physics* (Elsevier, Amsterdam).
5) W. Helfrich: *Z. Naturforsch.* **28c** (1973) 693.

Hydrodynamic Interprotein Interaction in a Biomembrane

S. Komura and T. Izuyama

Institute of Physics, College of Arts and Sciences, University of Tokyo,
Komaba 3-8-1, Meguro-ku, Tokyo 153, Japan

Abstract. A membrane of the lipid bilayer is regarded as a two-dimensional viscous and incompressible fluid coupled to the surrounding water. A long-range interaction arises between protein molecules moving in this two-dimensional fluid in a weak coupling limit.

1. Introduction

A biomembrane may be treated as a two-dimensional viscous and incompressible fluid. The lateral diffusion of protein molecules in the biomembrane is investigated. Here a protein molecule is represented by a hard disk.

The biomembrane layer is sandwiched by semi-infinite regions of water and the momentum of lipid molecules may leak into the surrounding water system. From this point of view, Saffman and Delbrück first obtained the mobility of the disk by taking into account both the thickness of the membrane and the viscosity of the outer water[1,2]. They predicted that the mobility of a protein inserted in the membrane depends very weakly on its size. The size dependence is only logarithmic. We stress that this result suggests the presence of a long-range interaction among membrane proteins.

In this report, we consider a simplified version[3] of the Saffman-Delbrück model, which is free from the Stokes paradox. The diffusion coefficient of protein molecules is obtained through the Einstein relation. Then we calculate the interprotein forces arising from the hydrodynamical origin.

2. Hydrodynamic Model with Momentum Decay

Let us start with the hydrodynamical equation

$$-\frac{1}{\rho}\text{grad } P + \nu\nabla^2\vec{v} - \frac{1}{\tau}\vec{v} = 0, \qquad \text{div } \vec{v} = 0, \tag{1}$$

where $\vec{v}$ is the velocity field of the two-dimensional fluid, ρ is the two-dimensional density, ν is the kinematical viscosity, grad, ∇^2 and div are the two-dimensional operators, P is the hydrostatic pressure, and τ is the relaxation time of the momentum of the lipid assembly.

Springer Proceedings in Physics, Vol. 52 Dynamics and Patterns in Complex Fluids
Editors: A. Onuki · K. Kawasaki

Consider a disk A with radius a moving with a constant velocity $-\vec{U}$. We take the Cartesian coordinates (x, y) with the origin fixed at the center of the disk and the x-axis along the direction $\vec{U} = (U, 0)$. The polar coordinates will be (r, θ). In this reference frame we write $\vec{v} = \vec{U} + \vec{u}$ and replace $-\vec{v}/\tau$ by $-\vec{u}/\tau$ in (1). The boundary conditions are $\vec{u} = 0$ for $r \to \infty$ and $\vec{v} = 0$ at $r = a$. An important parameter turns out to be $\kappa \equiv 1/\sqrt{\nu\tau}$.

In the weak coupling limit ($\kappa a \ll 1$), the diffusion coefficient is obtained by the Einstein relation as

$$D = \frac{k_B T}{4\mu\pi} \log \frac{1}{\kappa a}, \tag{2}$$

where $\mu = \rho\nu$. This agrees with Saffman-Delbrück's result. The pressure field around the disk is

$$P = \frac{\mu U}{\log \kappa a} \frac{\cos\theta}{r}. \tag{3}$$

Consider next two disks A and B, moving with velocities U and U', respectively. Let A be at the origin (0,0) and B be at $(x, y) = (r\cos\theta, r\sin\theta)$. For $\kappa r \gg 1$, the force acting on B is obtained as

$$F_x^{(B)} = C\pi a \frac{\cos 2\theta}{r^2}, \qquad F_y^{(B)} = C\pi a \frac{\sin 2\theta}{r^2}, \tag{4}$$

where $C = 2\mu U / \log \kappa a$. This is our main result. The hydrodynamic interaction is rather long-ranged. Details of the calculation will be published elsewhere.

3. Conclusion

It is remarkable that even in the presence of a momentum dissipation mechanism, the interprotein interaction is long-ranged as seen from (4). In the many protein system in the membrane, each protein does not move independently. The system exhibits collective motions leading to certain patterns.

References

1. P.G.Saffman and M.Delbrück: Proc. Natl. Acad. Sci. USA **72** (1975) 3111.
2. P.G.Saffman: J. Fluid. Mech. **73** (1976) 593.
3. Y.Y.Suzuki and T.Izuyama: J. Phys. Soc. Jpn. **58** (1989) 1104.

Ordering Process Below the Pre-transition Temperature in Multilamellar Lipid Solutions

S. Ueno, T. Takeda, S. Komura, and Y. Toyoshima

Faculty of Integrated Arts and Sciences,
Hiroshima University, Hiroshima 730, Japan

Abstract. Small-angle scattering studies of X-rays and neutrons on multilamellar lipid solutions quenched from the middle temperature phase to low temperature phase show a relaxation process of a simultaneous ordering of new structure and decaying of old structure having the same relaxation time, which depends on the quenched-in temperature.

1. Introduction

Since the physiological function of the biological membranes depends strongly on their structure, many studies have been made on the variation of the structure of lipid bilayers in aqueous solution with a change of temperature or addition of other substances. However, there still remain many problems to be solved. In excess water, phospholipids generally have a multilamellar structure which is an alternating stack of phospholipid bilayers and water. In aqueous dispersion of dipalmitoylphosphatidylcholine (DPPC) the L_{β}' phase exists below the pre-transition temperature T_{m1} of 33.5 °C, the P_{β}' phase between T_{m1} and the main transition temperature T_{m2} of 41.5 °C and the L_{α} phase above T_{m2}. We have studied the variation of the structure with temperature and the ordering process of the L_{β}' phase in aqueous dispersion of DPPC by means of small-angle X-ray scattering (SAXS) and small-angle neutron scattering (SANS). The wavelength of the neutrons and X-rays is about 4 Å and 1.542 Å, respectively. The lipid was dissolved in D_2O at 40wt% for SANS and in H_2O at 40wt% for SAXS.

Since the scattering intensity is given by a product of the intramembrane structure factor and the intermembrane structure factor, the scattering intensity of X-rays is different from that of neutrons because of the difference of the intramembrane structure factor. The results of a model calculation of the scattering intensity of X-rays and neutrons [1] suggest that X-ray diffraction has a high intensity for a scattering vector Q larger than Q_1, which is that for the 1st Bragg peak of the multilamellar structure, while neutron diffraction has a high intensity for Q smaller than Q_1.

2. Results

The results of SANS [2] show that there is a hysteresis in the scattering profile between heating and cooling processes. The relaxation process of the scattering intensity toward an equilibrium state was also found when the system was quenched from the middle temperature to the low temperature phase. In the SANS study, this relaxation process manifested itself in the decrease of diffuse scattering at a Q smaller than Q_1, while the intensity of the 1st Bragg peak remains the same.

On the other hand we observed in the SAXS study that the integral intensity of the 1st Bragg peak increases with time (Fig.1), which suggests that the L_{β}' phase is in the process of being ordered.

Figure 2 shows the temperature dependence of the relaxation time calculated from the time variation of the intensity of the diffuse scattering in SANS and that of the first Bragg peak in SAXS at the temperature after quenching.

Springer Proceedings in Physics, Vol. 52 **Dynamics and Patterns in Complex Fluids**
Editors: A. Onuki · K. Kawasaki

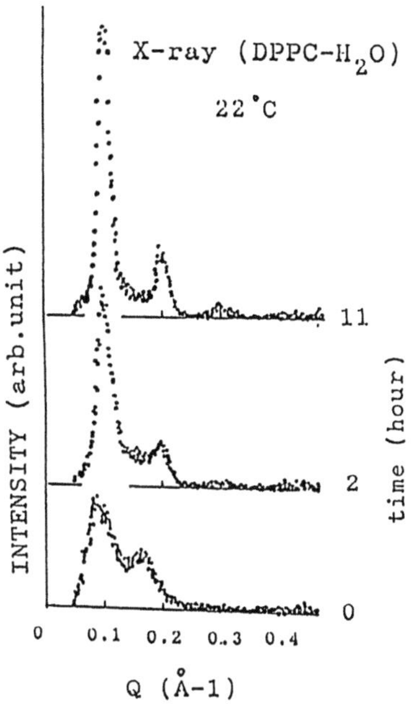

Fig. 1 Time variation of SAXS on DPPC in H_2O solution in the L_{β}' phase after quenching from P_{β}'.

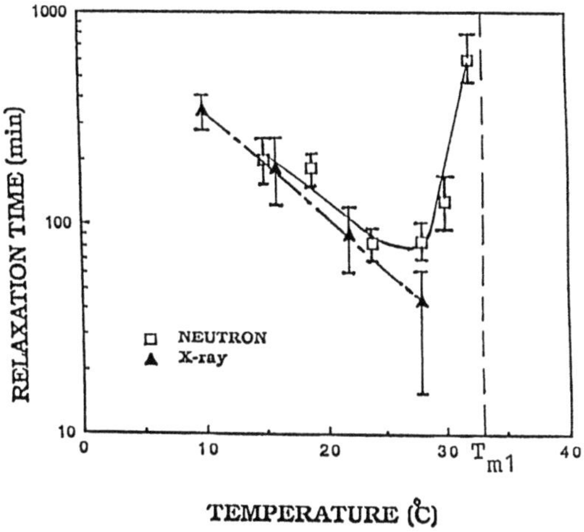

Fig. 2 Relaxation time of intensity change associated with structural change by means of SANS and SAXS as a function of temperature in L_{β}' phase after quenching from P_{β}'.

3.Discussion

Below 28°C the temperature dependence of the relaxation time obtained from the SAXS study essentially shows the same behavior as that from the SANS study. This result suggests that SANS measures a decaying process, while SAXS measures an ordering process at different ranges of Q in the same phenomenon. The temperature dependence of the relaxation time suggests that the diffusion coefficient of the DPPC molecules becomes smaller with decreasing temperature, while the critical slowing down near T_{m1} causes a slower ordering process near T_{m1}, giving rise to a minimum of the relaxation time at around 27°C.

We have tried a model calculation of the scattering intensity, assuming that the fluctuation of the difference of the repeat distance from the mean value d=62Å would be a gaussian distribution with a width γ. The calculated scattering profiles agree with those observed when we take the decreasing value of γ from 20 Å to 10 Å, as the time proceeds after the quench. This result suggests that the fluctuation of repeat distance is getting smaller after the quench and that the repeat distance gradually approaches a constant.

References

[1] T. Takeda, K. Akabori, Y. Toyoshima, S. Komura and Y. Takebe: Jpn. J. Appl. Phys. **26** (1987) 1791.

[2] T. Takeda, S. Ueno, S. Komura, Y. Toyoshima and K. Akabori : In "Dynamics of Ordering Processes in Condensed Matter", edited by S. Komura and H. Furukawa (Plenum Press, New York, 1988) p.543.

Anisotropic Dynamics in an Oriented Lipid Multibilayer Revealed by Vector EPR Spectroscopy

Y. Shimoyama and H. Watari

Physics Department, Hokkaido University of Education, Hakodate 040, Japan, and Department of Molecular Physiology, National Institute for Physiological Sciences, Okazaki 444, Japan

Abstract Vector electron paramagnetic resonance spectroscopy (V-EPR) was successfully applied to an oriented lipid multibilayer to explore the anisotropic dynamics of fluctuations in the multibilayer in the slow-motion domain. V-EPR analyzed accurately the slow anisotropic motions in the gel phase. Anisotropic structure in the correlation times of molecular fluctuation was found for the first time in the oriented lipid membranes, indicating that the anisotropy exists in both the magnitude and the time constant of the lipid motion. Axial rotation is more dominant than wobbling in the gel phase of multibilayer membranes.

Introduction

The spin labeling method with electron paramagnetic resonance (EPR) has been widely utilized to detect dynamic structure in lipid membranes [1]. The time scale of molecular motion in biomembranes ranges from 10^{-3} to 10^{-11} s. However, fluctuation in the gel phase of lipid membranes cannot be analyzed by the conventional EPR, because the time range (10^{-3} - 10^{-7} s) is much slower than the rigid limit [2]. Saturation transfer (ST) EPR has been developed to explore such slow motion of lipid ensembles in the gel phase [2]. Recently, we have shown that an integrated value of an ST-ESR signal changes in a sinusoidal manner and hence can be treated as a vector [3,4]. Using the vector EPR (VST-EPR), we could evaluate reliable rotational correlation times. In the present communication, we will report evidence of anisotropy in the correlation times of the lipid motion in oriented multibilayers revealed by the integrated values and further discuss the dynamic structure of the lipid membranes.

Experimental

Lipids used in this study were dimyristoyl phosphatidyl-choline (DMPC), and cholesterol which was added with 5 and 10 M% ratios. A spin label of 3-spiro-doxyl-5-cholestane (CSL) was mixed into the above mixture at 0.5 M%. An oriented multibilayer film was prepared by deposition of the mixed solution on a quartz plate (3 x 10 mm) and evaporation under a wet nitrogen gas stream at a temperature just above the gel-liquid crystalline transition point. All the VST-EPR spectra were measured at the second harmonic absorption display (V_2) modulated by 50 kHz and demodulated by 100 kHz. The high-power phase maximum method [cf. Ref. 4] was used to determine the 90 degree out of phase display (V_2'). Correlation times of molecular motion were determined using the calibration curve of the integrated value vs correlation time [5].

Results and Discussion

ST-EPR spectra of oriented multibilayers of DMPC containing 0, 5 and 10 M% cholesterol were recorded with the magnetic field perpen-

Springer Proceedings in Physics, Vol. 52 Dynamics and Patterns in Complex Fluids
Editors: A. Onuki · K. Kawasaki

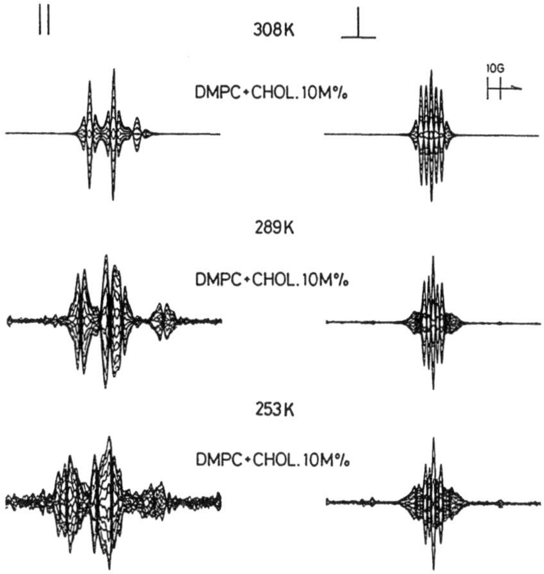

Fig. 1 V-EPR spectra DMPC-CSL multibilayer membranes recorded at second harmonics and in various phases.

dicular and parallel to the multibilayer film surface (Fig. 1). The integrated value was also shown at various magnetic fields. At high microwave power, however, the integrated spectra cannot be transformed to the V_1 signal and have a significant residue at the end of field sweep. For the DMPC multibilayer in the absence of cholesterol, approximately three phases and one phase transition were detected in each direction (Fig. 2A). The integrated values are almost

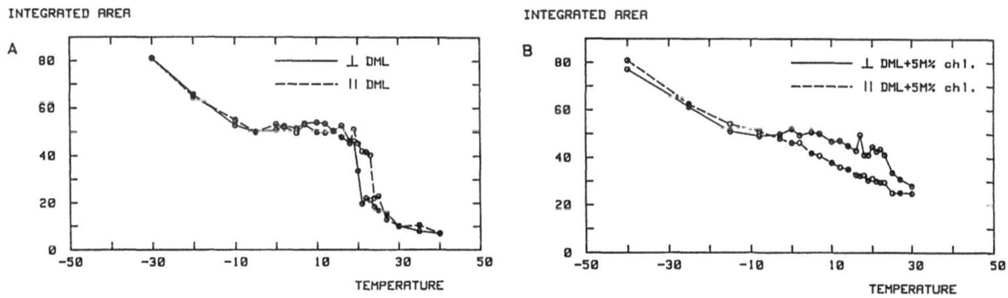

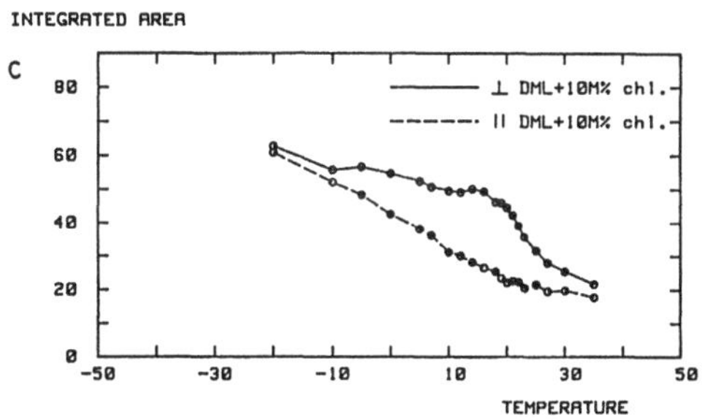

Fig. 2 Temperature variation of integrated value as evaluated by single integration of V_2.

identical at all phases except the phase transition region, i.e. 292 K to 296 K, where anisotropy was found. This indicates that the onset of tilt motion occurs prior to the long axis rotation [6]. In the case of 10 M% cholesterol, the integrated value of the perpendicular direction is larger than that of the parallel direction at temperatures of the gel phase (Fig. 2C). The maximum difference was observed at a temperature of the gel phase just below the phase transition region where the perpendicular value is two times larger than the parallel one. Such anisotropy tends to converge at both low and high temperature regions. Using the calibration curve, we determined the characteristic time of the wobbling motion of the lipid molecule in the oriented multibilayer to be 10^{-3} to 10^{-6} s and that of the axial rotation to be 10^{-6} to 10^{-9} s [5].

The temperature variations of the DMPC multibilayers both with and without the addition of cholesterol (Fig. 2B and C) indicate that an anisotropic structure exists in the correlation time of the lipid motion. This is clear evidence that the addition of cholesterol influences not only the molecular order but also the correlation time of the lipid ensemble. The variation of the integrated value shows that axial rotation is dominant over wobbling in the gel phase of multibilayer membranes.

Acknowledgment: This work was partly supported by a Grant-in-Aid (NO 61850001) of the Ministry of Education, Science and Culture of Japan.

References

[1]H. M. McConnell, in Spin Labeling; Theory and Applications, L. J. Berliner, Ed. (Academic Press, New York) p. 525 (1976).
[2]D. Thomas, L. Dalton and J. Hyde, J. Chem. Phys. **65**, 3006 (1976).
[3]Y. Shimoyama and H. Watari, Appl. Spectrosc. **39**, 170 (1985).
[4]Y. Shimoyama and H. Watari, J. Chem. Phys. **84**, 3688 (1986).
[5]Y. Shimoyama and H. Watari, Membranes **12**, 40 (1987).
[6]M. Janiak, D. Small and G. Shipley, Biochemistry **15**, 4575 (1976).

Part II

Gels

Swelling of Gels and Diffusion of Molecules

Yong Li and T. Tanaka

Department of Physics and Center for Material Science and Engineering,
Massachusetts Institute of Technology, Cambridge, MA 02139, USA

The kinetics of swelling and shrinking of gels is studied. A new relation, in addition to the differential equation developed by Tanaka and Fillmore, is formulated to solve the kinetics of gels having arbitrary shape. The gel kinetics is described as a combination of the collective diffusion with finite rate and immediate relaxation of shear deformation. The relation demonstrates the fundamental differences between the gel kinetics and the molecules diffusion process. The difference is a direct result of the existence of the shear modulus of the gel network system. Some interesting details of our theory are further discussed.

I. INTRODUCTION

The kinetics of gels is not only an interesting theoretical problem, but also a powerful probe which can be used to determine the mechanical and thermaldynamic properties of gels [1]. The volume relaxation of gels has been considered as a diffusion process known as the collective diffusion of polymer networks. The equation of motion of gel network in still solvent was first introduced by Tanaka, Hocker and Benedek [2] (THB) and later was used by Tanaka and Fillmore [3] (TF) to describe the swelling kinetics of spherical gels for the case of zero shear modulus, $\mu = 0$. The TF theory successfully explained the power law dependence of the relaxation time of gel on its size [3,4,5,6]. Peters and Candau further generalized the TF calculation to include $\mu \neq 0$ case [6] and studied the kinetics of long cylinder and large disc shaped gels [7]. We studied the same subject recently [1] and found that the TF theory is insufficient for gels with a non-spherical shape and the gel kinetics in general is not a pure diffusion process. In this article, we discuss in detail the difference between the gel swelling process and the molecules diffusion process. For the sake of completeness, some of the materials in reference [1] will appear here.

To proceed, let us first briefly review the TF theory. The equation of the network motion in still solvent is [2,8]

$$f\frac{\partial \mathbf{u}}{\partial t} = \nabla \cdot \tilde{\sigma}, \tag{1}$$

where $\mathbf{u}$ is the displacement vector of the network measured from the final equilibrium state and f is the friction coefficient between the network and the solvent. The stress

Springer Proceedings in Physics, Vol. 52 **Dynamics and Patterns in Complex Fluids**
Editors: A. Onuki · K. Kawasaki

tensor σ_{ik} and strain tensor u_{ik} are defined as

$$\sigma_{ik} = K\nabla \cdot \mathbf{u}\,\delta_{ik} + 2\mu\left(u_{ik} - \frac{1}{3}\nabla \cdot \mathbf{u}\,\delta_{ik}\right),$$

$$u_{ij} = \frac{1}{2}\left(\frac{\partial u_i}{\partial x_j} + \frac{\partial u_j}{\partial x_i}\right).$$

Using the above two equations, Eq. (1) can be written as

$$\frac{\partial \mathbf{u}}{\partial t} = D_0 \nabla^2 \mathbf{u} + \frac{K + \mu/3}{f} \nabla \times (\nabla \times \mathbf{u}), \qquad (2)$$

where $D_0 = (K + 4\mu/3)/f$ is the collective diffusion constant. In the cases of sphere, long cylinder, and large disc shaped gels, the last term is zero. In these cases, the above equation becomes a pure diffusion equation. Using this equation, Tanaka and Fillmore calculated the kinetics of spherical gels [1] with zero shear modulus:

$$u(a,t) = \frac{6\Delta}{\pi^2} \sum_n n^{-2} e^{-(n\pi/a)^2 D_0 t}, \qquad (3)$$

where Δ is defined as the total change of the radius of the gel and $\Delta = u(a, 0)$.

The insufficiency of TF kinetics theory for non-spherical gels can be demonstrated intuitively by comparing the diffusion of molecules, say ink, in water and the swelling of a gel. As shown in Fig.1a, imagine a long string of ink is placed in water at time zero. A pure diffusion process will occur and the ink string will become thicker and longer but the relative change of the length will be negligible compared to that of the diameter [9]. In the case of gel, however, experiments show that the relative change of the diameter and the length are the same (Fig.1b). This indicates that the gel swelling and shrinking processes are not pure diffusion processes.

This insufficiency is due to the existence of the shear modulus of the network. The nature of the shear modulus is to keep the system in shape, i.e., minimize the

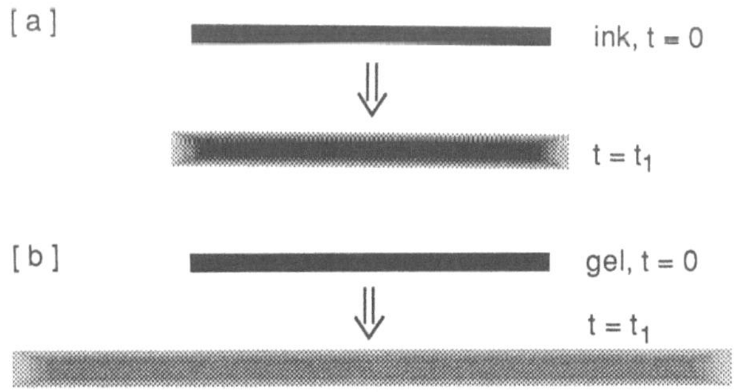

Figure 1. A schematic description of a diffusion process of ink molecules in water [a] and the gel swelling process [b]. For the former, the relative change of the length is negligible compared with that of the diameter. In the swelling of a long cylindrical gel, the relative change of the length and the diameter are the same.

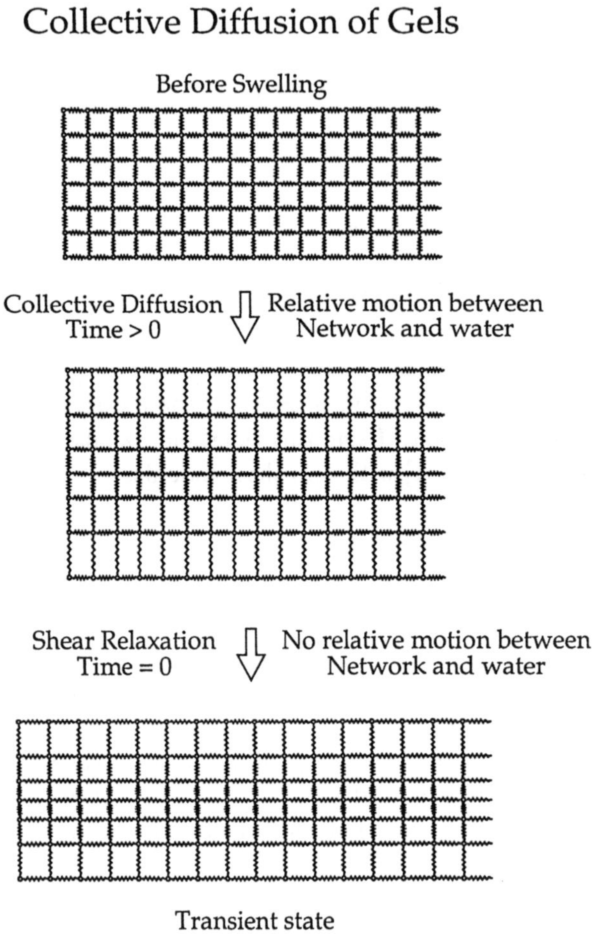

Figure 2. The swelling of a cylindrical gel network can be considered as a two-step process. The diffusion process makes the gel thicker but builds up the shear energy. Then the shear relaxation process will take place to minimize the shear energy. The second process is instantaneous and makes the swelling isotropic.

non-isotropic deformation. Since during a shear relaxation process there is no relative motion and hence no friction between gel network and solvent, the system can instantly adjust its shape to minimize the total shear energy. For a long cylindrical gel, any change in diameter is coupled to a change in length. This is essentially why the relative change of the diameter and the length are in proportion (see Fig.2).

In general, the total energy of a gel can be separated into a bulk energy and a shear energy [8]. The bulk energy of the system is related to the volume change, which is controlled by diffusion. The shear energy, on the other hand, can be minimized instantly by readjusting the shape of the gel. The shear energy of a gel of arbitrary shape is [8]

$$F_{sh} = \mu \int_V \left[\left(u_{xx} - \frac{T}{3} \right)^2 + \left(u_{yy} - \frac{T}{3} \right)^2 + \left(u_{zz} - \frac{T}{3} \right)^2 \right] dV \,, \tag{4}$$

where $T = (u_{xx} + u_{yy} + u_{zz})$ is the trace of the strain tensor u_{ik}. Without losing any generality, we have chosen the coordinate system to be the one in which u_{ik} is diagonalized. At any instance, this energy is minimized. As long as the shear modulus μ is not zero, the change of the total shear energy in response to any small pure shear deformation for every volume element within the gel should be zero,

$$\delta F_{sh} = 0 \,. \tag{5}$$

This equation plus the TF diffusion equation (2) should determine the kinetics of the gel swelling and shrinking processes. Every tiny diffusion process determined by TF equation (2) is coupled to (or corrected by) a small shear process governed by Eq. (5).

In the case of a sphere, Eq. (5) is a trivial relation because the shape of the sphere is maintained automatically by the diffusion process and the shear energy is already at its minimum with respect to its shape. Because of this, the calculations by TF and Peters and Candau in the case of spherical gels are valid [10].

For an infinitely long cylinder gel, u_{zz} is position independent. For any given u_{rr}, u_{zz} is determined by minimizing the total shear energy with respect to u_{zz}, i.e., $\partial F_{sh} / \partial u_{zz} = 0$. From Eq. (5) we have

$$u_{zz} = \frac{1}{A} \int \left[\frac{u_{rr} + u_{\phi\phi}}{2} \right] dA. \tag{6}$$

Therefore, the element u_{zz} is the average of the other diagonal elements. Substituting $u_{rr} = (\partial u_r / \partial r)$ and $u_{\phi\phi} = u_r / r$ into Eq. (6) and integrating by parts, we get

$$\frac{u_z(z,t)}{z} = \frac{u_r(a,t)}{a} \,. \tag{7}$$

Eq. (7) clearly indicates that the relative change of long cylindrical gels is isotropic, i.e., the swelling rate of a cylinder in the axial and radial directions are the same. This is exactly what we have observed experimentally.

The rest of this paper is organized as follows. In section II, we introduce the two-process approach and present in detail the theoretical treatment of the swelling kinetics of long cylindrical gels. The details and the relation between gel swelling and molecules diffusion will be discussed in section III.

II. KINETICS OF LONG CYLINDRICAL GELS

The kinetics of the swelling of a long cylindrical gel can be understood in the following manner. Let us consider an infinitesimally small swelling process as a combination of two consecutive small processes, as shown in Fig.3. The first process is a pure diffusion process described by Eq. (2) in which the diameter swells to a certain degree (by $\delta \bar{u}_r$) and the change of the length is negligible (zero for an infinitely long cylinder).

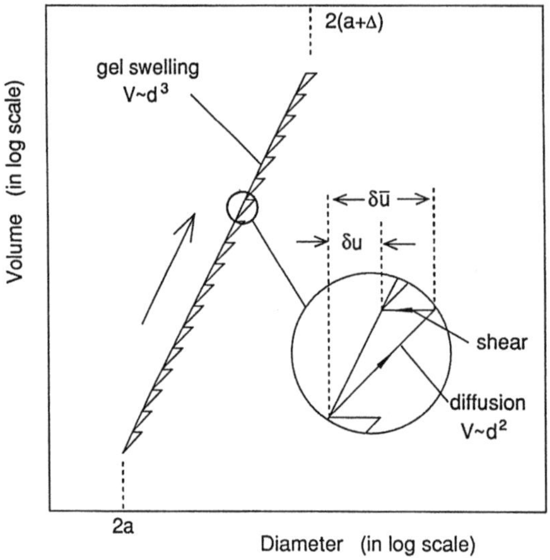

Figure 3. A schematic description of the two-process approach that describes the kinetics of a cylinder. In the figure, V is the total volume, d is the diameter. The initial diameter is 2a, and the final diameter is 2(a+Δ). The diffusion process changes the diameter by $\delta\bar{u}$, then the shear process reduces this value to δu.

During this process, the shear energy builds up. The second process is the shear relaxation process described by Eq. (5). In this process, the length of the gel will expand (by δu_z) and the diameter will shrink ($\delta\bar{u}_r$ becomes δu_r) to minimize the shear energy, with the volume kept constant. The consequence of the constant volume restriction is that there is no relative motion between the network and the solvent, hence no friction in the shear process. The shear process is therefore instantaneous. Now assume the change of the displacement vectors (u_r, u_z) from time t to $t+\delta t$ are $(\delta\bar{u}_r, 0)$ after the diffusion process and $(\delta u_r, \delta u_z)$ after the shear process, as indicated in Fig.3. The changes $(\delta\bar{u}_r, 0)$ are intermediate quantities. Notice that for infinitely long gels there is no change in u_z during the diffusion process.

The two changes δu_r and $\delta\bar{u}_r$ are related through the shear relaxation process. Since the volume of every volume element is preserved in this process, we have

$$(r-u_r-\delta u_r)^2(z-u_z-\delta u_z) = (r-u_r-\delta\bar{u}_r)^2(z-u_z)\,,$$

or

$$\frac{\delta u_r}{\delta t} + \frac{1}{2}\frac{r-u_r}{z-u_z}\frac{\delta u_z}{\delta t} = \frac{\delta\bar{u}_r}{\delta t}\,.$$

We have used the fact that δu_r and $\delta\bar{u}_r$ are the changes that occurred in time δt. Define

$$\lambda \equiv \frac{1}{2}\left(\frac{1}{z-u_z}\frac{\partial u_z}{\partial t}\right) \Big/ \left(\frac{1}{r-u_r}\frac{\partial u_r}{\partial t}\right) \tag{8}$$

$$\approx \frac{r}{2z}\left(\frac{\partial u_z}{\partial t}\right) \Big/ \left(\frac{\partial u_r}{\partial t}\right),$$

then

$$\frac{\delta u_r}{\delta t} = \frac{1}{1+\lambda}\frac{\delta \bar{u}_r}{\delta t}. \tag{9}$$

It is clear from this equation that the effect of the shear process is to reduce the speed of the diffusion process by $1/(1+\lambda)$. We will show later that the apparent local collective diffusion constant is also related with this factor.

In cylindrical coordinates, the diffusion equation (2) becomes

$$\frac{1}{D_0}\frac{\partial \bar{u}_r}{\partial t} = \frac{\partial^2 \bar{u}_r}{\partial r^2} + \frac{1}{r}\frac{\partial \bar{u}_r}{\partial r} - \frac{\bar{u}_r}{r^2}. \tag{10}$$

We have used the notation $\bar{u}_r$ to emphasize that the solution of this equation is not the final solution of the problem. The general solution of this equation is

$$\bar{u}_r(r,t) = \Delta \sum_n \bar{A}_n J_1(q_n r) e^{-D_0 q_n^2 t}, \tag{11}$$

with the eigenvalues q_n to be determined by the boundary condition.

We shall seek a final solution of the form

$$u_r(r,t) = \Delta \sum_n A_n J_1(q_n r) e^{-D_e q_n^2 t}, \tag{12}$$

where $D_e = D_e(r,t)$ is a function of r and t. This solution must satisfy both the initial and boundary conditions. Also, it must be a solution of the two-process infinitesimal operator which is the combination of the pure diffusion and pure shear relaxation process.

The diffusion process at time t gives

$$\bar{u}_r(r,t+\delta t) = \Delta \sum_n \bar{A}_n(t) J_1(q_n r) e^{-D_0 q_n^2 \delta t}, \tag{13}$$

with $\bar{A}_n(t)$ determined by the initial condition of the infinitesimal process, which is $u_r(r,t)$ in Eq. (12). Therefore

$$\begin{aligned} \bar{u}_r(r,t+\delta t) &= \Delta \sum_n A_n J_1(q_n r) c^{-D_e q_n^2 t} c^{-D_0 q_n^2 \delta t} \\ &= u_r(r,t) + \delta \bar{u}_r(r,t). \end{aligned} \tag{14}$$

In time δt, the changes of $\bar{u}_r$ and u_r are

$$\delta \bar{u}_r = -\Delta \sum_n D_0 q_n^2 A_n J_1(q_n r) e^{-D_e q_n^2 t}\,\delta t, \tag{15}$$

$$\delta u_r = -\Delta \sum_n \left(\frac{\partial D_e t}{\partial t}\right) q_n^2 A_n J_1(q_n r) e^{-D_e q_n^2 t}\,\delta t. \tag{16}$$

Using Eq. (9), we have

$$\frac{\partial (D_e t)}{\partial t} = \frac{D_0}{1+\lambda}; \tag{17}$$

or

$$D_e(r,t) = \frac{1}{t}\int_0^t \frac{D_0}{1+\lambda(r,t)}dt\,. \tag{18}$$

From this equation, we find that due to the reduction effect of the shear process, the apparent collective diffusion constant D_e is smaller than the value of D_0, which can be measured directly from the kinetics of spherical gels. Using the result of Eq. (7), we get

$$u_z(z,t) = \frac{\Delta}{a} z \sum_n A_n J_1(\alpha_n) e^{-D_{e0} q_n^2 t}\,. \tag{19}$$

Where $D_{e0} = D_e(a,t)$ is time independent and will become clear shortly. The eigenvalues $\alpha_n = q_n a$ are determined by the boundary condition

$$\sigma_{rr}(a,t) = M\left[\frac{\partial u_r}{\partial r} + (1-2R)(\frac{u_r}{a} + u_{zz})\right] = 0, \tag{20}$$

where $M = (K + 4\mu/3)$ is the longitudinal modulus and $R = \mu/M$ is the ratio of the shear modulus to the longitudinal modulus. Notice that R varies from 0 ($\mu/K = 0$) to 0.75 ($\mu/K = \infty$). Using Eq. (12) and Eq. (19), the boundary condition becomes

$$R = \frac{1}{4}\left(1 + \frac{\alpha_n J_0(\alpha_n)}{J_1(\alpha_n)}\right)\,. \tag{21}$$

The coefficients A_n are determined by the initial condition $u_r(r,0) = r\Delta/a$,

$$A_n = \frac{2(3-4R)}{\alpha_n^2 - (4R-1)(3-4R)} \frac{1}{J_1(\alpha_n)} \tag{22}$$

$$\equiv B_n \frac{1}{J_1(\alpha_n)}, \tag{23}$$

where B_n, which is A_n times the corresponding eigenfunction evaluated at the boundary, is defined for future convenience. The parameter $\lambda = \lambda(r,t)$ is

$$\lambda = \frac{1+\lambda}{1+\lambda_0} \frac{r}{2a} \frac{\sum_n A_n q_n^2 J_1(\alpha_n) e^{-D_{e0} q_n^2 t}}{\sum_n A_n q_n^2 J_1(q_n r) e^{-D_e q_n^2 t}}, \tag{24}$$

or,

$$\frac{\lambda}{1+\lambda}(1+\lambda_0) = \frac{r}{2a}\frac{J_1(\alpha_1)}{J_1(\alpha_1 r/a)} + O\left(e^{-D_e q_2^2 t}\right)\,. \tag{25}$$

Where $\lambda_0 = \lambda(a,t)$. Thus the apparent diffusion constant D_e is position and time dependent. The leading order term of D_e for $R = 0$ is plotted as a function of radial position in Fig.4. As R increases, λ decreases and approaches the constant 1/2. In the case of $R \to 3/4$, $\lambda = 1/2$ exactly. It is also interesting to notice from Eq.(24) that at $r = a$, λ and D_e are independent of time t and normalized shear modulus R:

$$\lambda(a,t) = \lambda_0 = \frac{1}{2}, \tag{26}$$

$$D_e(a,t) = D_{e0} = \frac{2}{3}D_0\,. \tag{27}$$

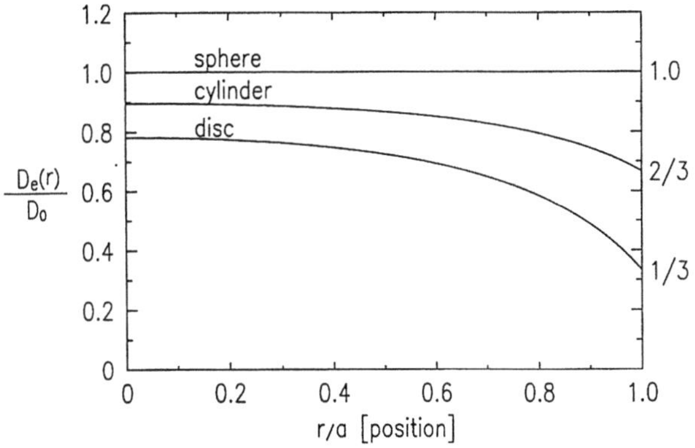

Figure 4. The position dependence of the effective collective diffusion constant normalized by the collective diffusion constant of spherical gels, $D_0 = (K+4\mu/3)/f$. At the boundary, the values for sphere, cylinder, and disc are 1, 2/3, and 1/3 respectively.

Therefore in an experiment where the shape and size of the gel are the quantities measured, the observed effective diffusion constant is time independent. The kinetics is described by pure diffusion in both r and z direction:

$$u_r(a,t) = \Delta \sum_n B_n \, e^{-t/\tau_n} , \tag{28}$$

$$u_z(z,t) = \Delta \frac{z}{a} \sum_n B_n \, e^{-t/\tau_n} , \tag{29}$$

where B_n is defined by Eq. (23) and $\tau_n = a^2/D_e\alpha_n^2$. To emphasize, we would like to point out that in the above two equations, the τ_n are independent of time t. A. Onuki [11] pointed out recently that τ_1 diverges as $\mu/K \to \infty$. This result can also be derived by using Eq.(21). In the case of $R \to 3/4$ ($\mu/K \to \infty$), the relaxation time $\tau_1 \sim 1/(3/4 - R)$ goes to infinity and all the B_n's go to zero except B_0, which goes to unity. So we have

$$u_r(r,t) \approx \Delta \frac{r}{a} e^{-t/\tau} , \tag{30}$$

$$u_z(z,t) \approx \Delta \frac{z}{a} e^{-t/\tau} . \tag{31}$$

This indicates that the larger the ratio μ/K, the smaller the contribution of the higher order terms and the better the single exponential approximation to the solution.

The above are reasonable results because when $\mu \to \infty$, the cost in shear energy of the diffusion process is very high, so the diffusion process will be slowed.

The treatment of a disc gel is similar to that of a cylinder gel , we will not go into it here. The theory described above has been confirmed excellently by experiments (see Table 1).

Table 1: Kinetic behavior of gels with different geometry. The last two columns are experimental results. The two numbers in each of the theoretical columns correspond to $\mu/K=0$ and $\mu/K=\infty$. The two values of τ_1 for the cylinder in column 5 represent the diameter and the length measurements (for details, see reference [1]).

	Theory				Experiment	
	α_1	B_1	D_e/D_0	$\tau_1/\tau_{1,sph}$	$\tau_{1,exp}(min)$	$\tau_1/\tau_{1,sph}$
sphere	$\pi\sim 0$	$0.608\sim 1$	1	1	39 ± 8	1
cylinder	$2.74\sim 0$	$0.573\sim 1$	2/3	$2.0\sim 1.9$	$(66, 65)\pm 8$	1.7 ± 0.3
disc	$2.29\sim 0$	$0.533\sim 1$	1/3	$5.7\sim 5.0$	215 ± 6	5.5 ± 0.9

III. Discussion

Following the definition of $\mathbf{u}(\mathbf{r},t)$, the actual position of a volume element with the final location (r,θ,z) is (r',θ,z') ,

$$r'(r,t) = r - u_r(r,t)\,, \tag{32}$$

$$z'(z,t) = z - u_z(z,t)\,. \tag{33}$$

The local swelling ratio in the axial and radial directions are $\partial z'/\partial z$ and $\partial r'/\partial r$, respectively. Using $u_z(z,t)$ and $u_r(r,t)$ obtained in the last section, we find

$$\left.\frac{\partial r'}{\partial r}\right|_{r=0} < \frac{\partial z'}{\partial z} < \left.\frac{\partial r'}{\partial r}\right|_{r=a}\,. \tag{34}$$

This means that the axial swelling ratio is larger than the radial swelling ratio near the center but smaller than that near the surface. In other words, the volume elements are elongated along the axial direction near the center but are elongated along the radial direction near the surface. This aspect is shown in the intuitive picture of the process indicated by Fig.2 and should be reflected in the time dependence of the effective collective diffusion constant $D_e(r,t)$.

To summarize, we find that the two most distinctive features of gel swelling process compared with a diffusion process are

1. *The swelling or shrinking rate, defined as the relative change from the initial value, is isotropic, i.e.,*

$$\frac{u_z(z,t)}{z} = \frac{u_r(a,t)}{a}\,.$$

2. *The effective collective diffusion coefficient D_e depends on both time and position. However, the value of D_e at the boundary is time independent.*

The first feature is totally unexpected from an ordinary diffusion process. It is a direct result of the existence of the shear modulus of the network. The second feature makes the process more complicated to analyze. According to this, measurements of the kinetics from different radial positions will yield different effective diffusion constants.

However, if we only measure the diameter of the gel, the second feature does not cause any difficulty.

It is clear now that the TF theory can only be used to describe a network system with absolutely no shear modulus, or spherical gels where shear energy minimization condition is always satisfied by the diffusion process. A network without shear elasticity adopts the shape of its container and should be considered as a solution rather than a gel. A gel with finite shear modulus, no matter how small the modulus is, behaves totally differently from the system with no shear modulus. The results in the $\mu \rightarrow 0$ and $\mu = 0$ are, therefore, totally different.

Consider the diffusion of molecules which, at time zero, are confined in a long cylinder of radius a. The initial concentration of the molecules inside the cylinder C_0 is uniform, and higher than that of the outside, which is a constant (well stirred) and equal to C_1. The surface of the cylinder allows the molecules to pass through with the rate proportional to the concentration difference between inside and outside (evaporation law). The diffusion process inside the cylinder can be described by the following equations,

$$\frac{1}{D}\frac{\partial C}{\partial t} = \frac{\partial^2 C}{\partial r^2} + \frac{1}{r}\frac{\partial C}{\partial r} - g(r)\,, \tag{35}$$

$$-D\frac{\partial C(a)}{\partial r} = \alpha(C(a) - C_1)\,, \tag{36}$$

$$C(r,0) = C_0\,, \tag{37}$$

here C is the concentration of the molecules inside the cylinder, $C(a)$ is the concentration at the surface, and α is the evaporation constant. The term $g(r)$ corresponds to the annihilation of the molecules. Compare this problem with the radial swelling of a cylindrical gel, we find that they are similar if we set $g(r) = C(r)/r^2$. The evaporation constant α corresponds to $(1-2R)/D$. The solution to this problem is

$$C(r,t) = C_1 + (C_0 - C_1)\sum_n A_n J_1(q_n r) e^{-Dq_n^2 t}\,. \tag{38}$$

By comparing this result with eq (12), we find that they look the same except in the case of gel, the effective diffusion constant D_e depends on both time and radial position. But the behavior of $u_r(a,t)$ is identical to $C(a,t)$ provided the diffusion constant of the molecules D equal to $D_e(a)$ of the gel. In the case of gel, it is the displacement field that relaxes. However, in the case of molecules the density profile relaxes.

The annihilation term $g(r)$ is necessary to draw the analogy. The term appears in the gel swelling case because $\mathbf{u}(\mathbf{r},t)$ is a vector field. This term does not appear when we make the analogy between the swelling of a disc gel and the diffusion of molecules in a disc container. The phenomena are essentially one-dimensional in this case.

Finally, we would like to point out that we have tried to calculate $D_e(r,t)$ as a function of t for fixed r. However, we find that the simple iteration method we used did not converge [11].

ACKNOWLEDGMENTS

The authors thank Terence Hwa, Kevin Wasserman, and Akira Onuki for valuable discussions. This work was supported by NSF, DMR87-19217.

References

[1] Y. Li and T. Tanaka, J. Chem. Phys. (to be published).

[2] T. Tanaka, L. Hocker and G. Benedek, J. Chem. Phys. **59**(9), 5151(1973).

[3] T. Tanaka and D. Fillmore, J. Chem. Phys. **70**(3), 1214(1979).

[4] T. Tanaka, E. Sato, Y. Hirokawa, S. Hirotsu and J. Peetermans, Phys. Rev. Lett. **55**(22), 2455(1985).

[5] E. Sato-Matsuo and T. Tanaka, J. Chem. Phys. **89**(3), 1695(1988).

[6] A. Peters and S. J. Candau, Macromolecules **19**(7), 1952(1986).

[7] A. Peters and S. J. Candau, Macromolecules **21**(7), 2278(1988).

[8] L. D. Landau and E. M. Lifshitz, *Theory of Elasticity*, 1986 Pergamon Press, Oxford.

[9] The boundary of the ink string can be defined as the density profile of the ink molecules at a certain value. For instance, we can define the radius of the string r as the value at which the relative density $\rho(r)/\rho(0) = 0.1$.

[10] Due to a mistake made in the initial condition, the coefficients A_n given by Eq. (6) in [6] are not correct. The correct answer should be

$$A_n = 4\frac{(\alpha_n^2 - 3)\sin\alpha_n + 3\alpha_n\cos\alpha_n}{2\alpha_n^2 + \alpha_n\sin(2\alpha_n) - 4\sin^2\alpha_n}\ .$$

The boundary condition of a spherical gel is $R = \alpha_n^2/(4 - 4\alpha_n\cot\alpha_n)$.

[11] Akira Onuki, Phys. Rev. A **38**(4), 2192(1988).

[12] This does not mean that there is no solution to this problem. Readers can verify easily that for equation $x^2 + x = 1$, iteration $x_{n+1} = 1 - x_n^2$ does not converge (but $x_{n+1} = \pm\sqrt{1 - x_n}$ does).

Phase Transition of D.N.A. Gel and Reentrant Phenomena in Acrylamide and Gelatin Gels

E. Takushi[1] *and T. Tanaka*[2]

[1]Physics Department, Ryukyus University, Okinawa 903-01, Japan
[2]Physics Department, Massachusetts Institute of Technology, Cambridge, MA 02139, USA

In 1987 Amiya and Tanaka demonstrated the universality of the first order phase transition in chemically cross-linked networks of natural polymer gels in acetone-water mixtures.[1,2] We here present the first observations of volume phase transitions of D.N.A. gels and reentrant transitions of acrylamide and gelatin gels in mixtures of ethanol-water at 26℃. D.N.A. in water (20%) is cross-linked by using EGDE(50% of D.N.A.) at pH 10-7 at 55℃ for 3 hours after 4 hours stirring. The volume change at the discrete transition points are V/V =5-56 times at 70-80% ethanol concentrations (Fig.1).

Reentrant volume phase transitions of acrylamide gel (ionized by sodium acrylate) are observed at 21% (V/V =15) and 71% (V/V =16) ethanol concentrations at 26℃ (Fig.2) and furthermore, gelatin polymer in water(10%) is cross linked by using EGDE in the amount of 10% of gelatin polymer at 60℃,pH 11 for 15 hours. Volume phase transitions of gelatin gel are also obtained at 70% and 90% (V/V =9.8 and 2.7 respectively) at 26℃.(Fig.3)

In order to interpret the reentrant phenomenon of acrylamide and gelatin gels we use the mean field theory of phase transition of ionic gels.[2] The osmotic pressure is described by the Flory-Huggins' formula,

$$\Pi = NkT/v[\phi + \ln(1-\phi) + (\Delta F/2kT)\phi^2] + \nu kT[(\phi/2\phi_0) - (\phi/\phi_0)^{1/3}] + f\nu kT(\phi/\phi_0). \quad (1)$$

The osmotic pressure inside the gel must be equal to that outside for the gel to be in equilibrium with the surrounding solvent. The requirement of zero

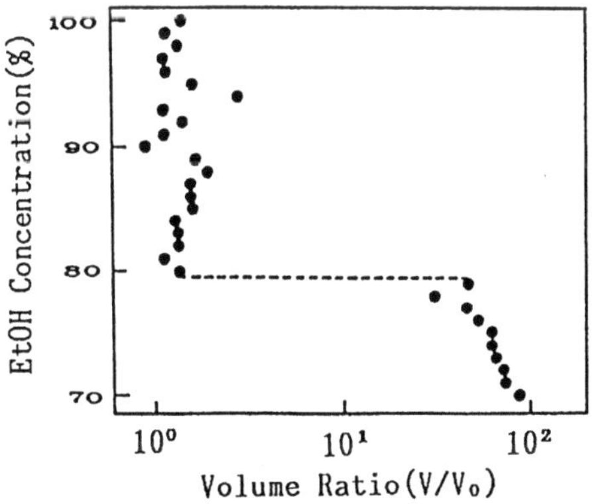

Fig.1.Swelling ratio of DNA gels in the ethanol-water mixtures at 26℃.

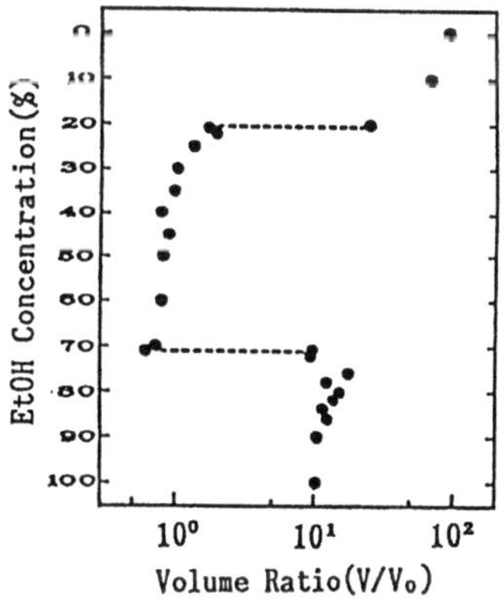

Fig.2.Reentrant swelling ratio of acrylamide gels in the ethanol-water mixtures at 26℃.

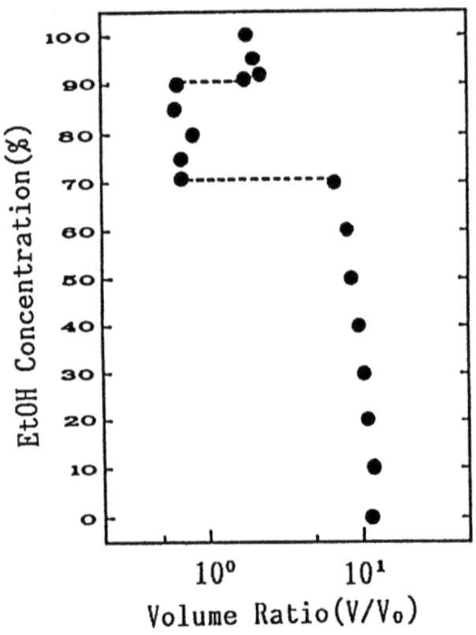

Fig.3.Reentrant swelling ratio of gelatin gels in the ethanol water mixtures at 26℃.

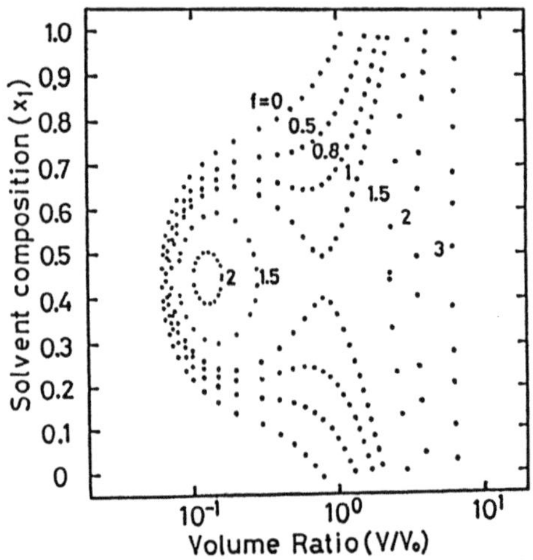

Fig.4.Theoretical relationship between gel volume,ionization(f) and solvent composition(X_1) calculated using the parameters $\Delta F_1/kT=0.4$, $\Delta F_2/kT=0.8$ and $\Delta F_{12}=-3.8$.

osmotic pressure difference is equivalent to minimizing the free energy of the gel F, since $\Pi=\partial F/\partial V$. Here V is the volume of the gel. From eq.(1), this condition is expressed as

$$\tau=1-(\Delta F/kT)=-(\nu v/N\phi^2)[(2f+1)(\phi/\phi_0)-2(\phi/\phi_0)^{1/3}] +1+(2/\phi)+2\ln(1-\phi)/\phi^2 \,. \tag{2}$$

Next we consider the dependence of τ on the composition of a mixed solvent The excess free energy ΔF may be described as a function of the solvent composition[3],

$$\Delta F=X_1\,\Delta F_1+X_2\,\Delta F_2-X_1\,X_2\,\Delta F_{12}\,, \tag{3}$$

where X_1 and X_2 ($X_1+X_2=1$) are the molar fraction of respective solvents 1 and 2, in the mixture. ΔF_1, ΔF_2 and ΔF_{12} are parameters defining the free energy of the association between solvent 1 and polymer segment, solvent 2 and polymer segment, and solvent 1 and solvent 2 respectively.

As a conclusion, (1) there exists a first order volume phase transition in DNA gel in the mixtures of EtOH and water, and (2) reentrant phenomena, such as acrylamide and gelatin gels, arise from the interaction between ethanol X_1 and water X_2 which is enhanced by the presence of polymer networks. More quantitatively,the interaction parameter ΔF_{12} should satisfy the condition $\Delta F_{12}<-|\Delta F_1-\Delta F_2|$ and the calculated results of solvent composition X_1 against volume ratio (V/V_0) are shown in Fig.4.

References

(1) T.Amiya & T.Tanaka, Macromolecules. 20, 1162 (1987).
(2) T.Tanaka, *et al.,* Phys.Rev.Lett. 45, 1636 (1980).
(3) T.Amiya, *et al.,* J.Chem.Phys. 86, 2375 (1987).

Spongelike Domain Structure in a Two-Dimensional Model Gel Undergoing the Volume Phase Transition

K. Sekimoto, N. Suematsu, and K. Kawasaki

Department of Physics, Kyushu University, Fukuoka 812, Japan

Abstract. We have studied the numerical model of a two-dimensional gel exhibiting a volumetric phase separation between the swollen and the shrunken phases. We have found a peculiar percolating (spongelike) structure of the shrunken phase domain.

Many polymer gels are known to exhibit the so-called first-order volume phase transition [1]. The primary order parameter of this transition is the swelling ratio s, or the ratio of the volume occupied by the gel material in question to the volume occupied by the same material in a certain isotropic and homogeneous reference state. In contrast to the liquid-gas phase transitions, the spatial phase coexistence in gels at the volume phase transition is subject to the constraint that the connectivity of polymer network in gels be unchanged under arbitrary deformation. This constraint inevitably produces shear strain, and therefore shear stress in the gel material when the swollen and the shrunken phase regions coexist.

To see how this constraint affects the morphology of the domains of these two phases, we have constructed a two-dimensional model of a gel [2] on the basis of the free energy F similar to the Flory model [3,4] having the following functional form:

$$F = \int d^2\mathbf{x}[f(s) + (\nu_0/2)(\lambda_1^2 + \lambda_2^2)],$$

where $d^2\mathbf{x}$ is the volume (area) element of the gel in its reference state. λ_1 and λ_2 are the principal stretching ratios of the volume element of the gel measured in reference to the reference state. Thus the swelling ratio is $s = \lambda_1\lambda_2$. The first part of the integrand, $f(s)$, is the contribution to the free energy due to dilation, and $f(s)$ has the two minima which correspond to the swollen phase and the shrunken phase. The remaining part in the integrand contains the elastic energy due to shear strain, for which $\lambda_1 \neq \lambda_2$, where ν_0 is a constant [2]. We have discretized the model material into many connected elements using the idea based on the finite element methods, and numerically sought the coexisting-phase states, which correspond to the local minima of the free energy F, by relaxing the model material from some initial states so as to lower the value of F.

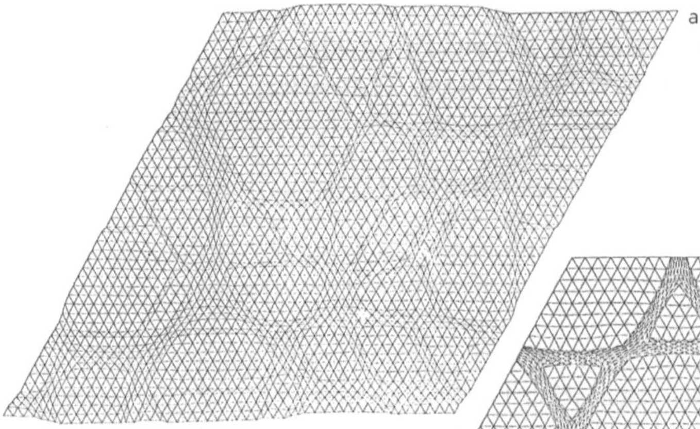

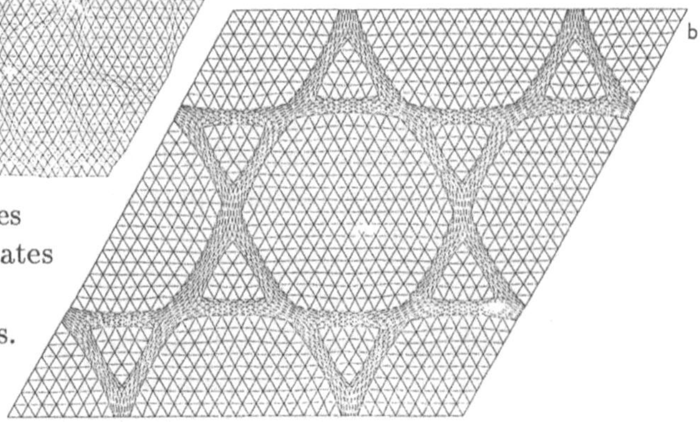

Fig.1 Coexisting-phase states obtained from the initial states with (a) random and (b) periodic weak distortions.

In Fig.1 we show three typical results of the coexisting-phase states. The triangular elements in Figs. 1(a) and 1(b) represent the deformation of material elements which were regular triangles of the same size in the reference state. Figure 1(a) shows the coexisting state relaxed from a certain random initial state. One can see the percolation of the shrunken phase domain. Figure 1(b) shows a nearly periodic coexisting state in accord with the periodic boundary condition which we have imposed. If we apply a pure shear to the system in the state shown in Fig.1(a), the domain boundaries between the two phases become aligned along the stretching direction [2]. These results demonstrate the qualitative difference from the phase coexistence in fluids, where the minority phase exists as more or less spherical domains. The recent experimental observation of NIPA gel [5] also shows spongelike structure. The corresponding numerical modelling in three dimensions is now under way.

References

1. T. Tanaka, these proceedings.
2. For details see K. Sekimoto and K. Kawasaki, in *Space-Time Organization in Macromolecular Fluids*, edited by F. Tanaka, M. Doi, T. Ohta, Springer Ser. Chem. Phys., Vol. 51 (Springer, Berlin, Heidelberg 1989). See also K. Sekimoto, N. Suematu and K. Kawasaki, Phys. Rev. A**39**, 4912 (1989).
3. P. J. Flory, *Principles of Polymer Chemistry* (Cornell Univ. Press, Ithaca, 1953)
4. A. Onuki, J. Phys. Soc. Jpn. **57**, 699 (1988).
5. S. Hirotsu and A. Kaneki, in *Dynamics of Ordering Processes in Condensed Matter*, edited by S. Komura and H. Furukawa (Plenum, New York, 1988),p.451.

Ordered Structures Resulting from Photo-cross-links of Binary Polymer Blends Undergoing Phase Separation

Q. Tran-Cong, T. Nagaki, M. Imura, O. Yano, and T. Soen

Department of Polymer Science and Engineering,
Kyoto Institute of Technology, Matsugasaki, Kyoto 606, Japan

1. Introduction

The formation of the highly ordered structures in the reaction-diffusion system driven far from equilibrium has been the subject of numerous theoretical and experimental studies in the last decade [1]. These phenomena have been observed in a wide class of physical systems where chemical reactions take place under non-equilibrium thermodynamic conditions[2]. In this work, the formation of the ordered structures resulting from the instability caused by U.V. light irradiation of a binary polymer blend is reported.

2. Experimental Section

Polymer blends used in this work are the mixtures of polystyrene (PS, Mw = 140K, Mw/Mn = 1,9) and poly(2-chlorostyrene) (P2CS, Mw = 84K, Mw/Mn = 2.2). Anthracene as a photo-reactive group was chemically labeled on PS chains (PSA)[3]. Upon irradiation with U.V. light, anthracenes on PS chains react and form a network as schematically shown in figure 1. The phase diagram of PSA/P2CS blends used in this experiment is of the LCST (lower critical solution temperature) type. PSA/P2CS blends with the composition of 40/60 (PSA/P2CS) were jumped from the one-phase region (166.6^{0}C) into the two phase region (185^{0}C and 192 8^{0}C). Subsequently, U.V. light (351nm) from a XeF excimer laser(Lambda Physik, 15ns in pulse width) was impinged on the blends during the time-evolution of the spinodal decomposition process by following the experimental procedure described previously[4]. After irradiation, PSA/P2CS blends were solidified by cooling to room temperature. The morphology of irradiated blends was observed under a phase-contrast optical microscope. For a PSA/P2CS blend submitted to a shallow jump, the irradiated structures were observed by using a light scattering photometer since these structures are beyond the resolution of optical microscopy.

3. Results and Discussion

As shown in figure 2, for a PSA/P2CS blend submitted to a deep jump (192.8^{0}C) into the spinodal region, the interconnecting structures were formed by 5min of irradiation which starts at 15min after the temperature jump (T-jump) Without irradiation, as shown in figure 3, the two-phase structures are clearly seen under optical microscope for the same PSA/P2CS blend. These results indicate that upon irradiation by U.V. light, due to the cross-link reactions of anthracene, the networks of PS chains were formed and restricted the diffusion of P2CS chains during the phase decomposition process. As a consequence, the interconnecting structures generated by the spinodal process were frozen

On the other hand, for a PSA/P2CS blend submitted to a shallow jump (185^{0}C), upon irradiation in 5min at 5min after the T-jump, the morphology of the irradiated blend could not be observed by a phase-contrast optical microscope. Therefore a light scattering photometer was used. Surprisingly, the light scattering profile shows higher ordered peaks as shown in figure 4 This scattering profile also changes with annealing times after irradiation. Taking into account the non-equilibrium thermodynamic conditions caused by irradiation, especially by the gradient of radiation between the two surfaces of the sample, at this moment we think that this experimental result strongly suggests the existence of a new kind

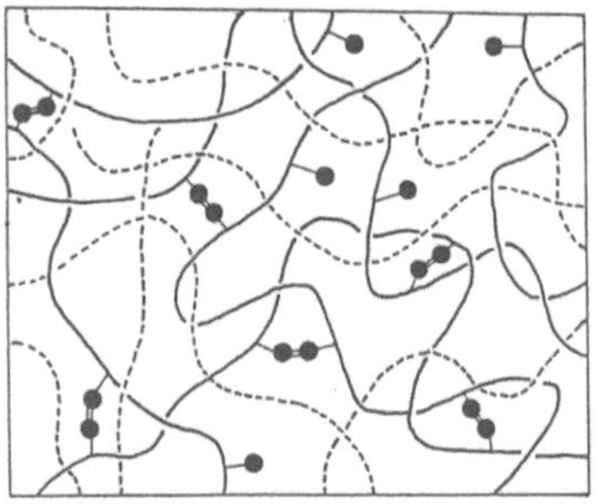

Fig. 1. Schematical representation of photo-cross-links: (——), PSA; (---), P2CS; (●): anthracene.

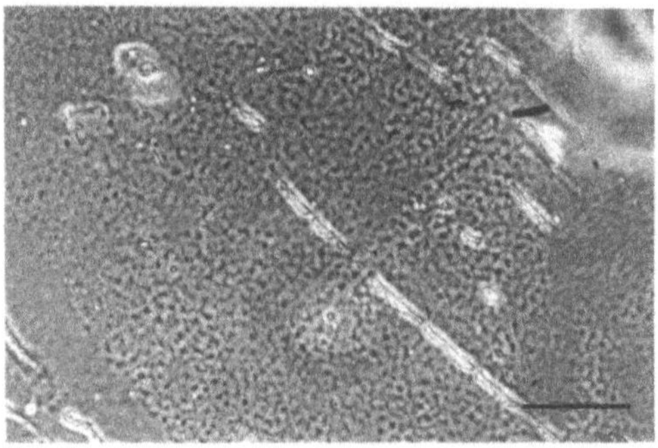

Fig. 2. Optical micrograph of a PSA/P2CS blend irradiated in 5min at 15min after the T-jump. The scale is 20μm.

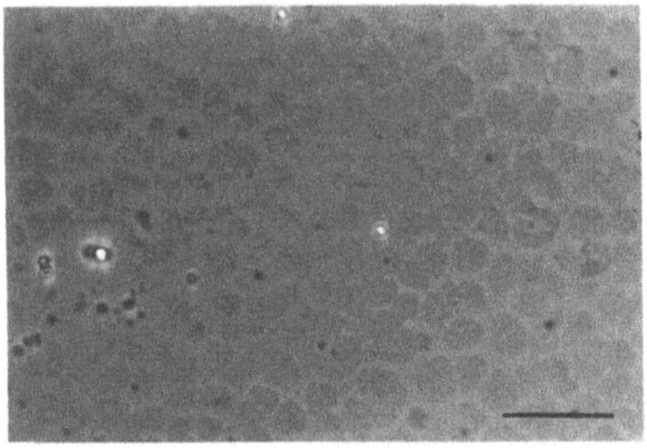

Fig. 3. Optical micrograph of a PSA/P2CS blend submitted to the T-jump without irradiation,

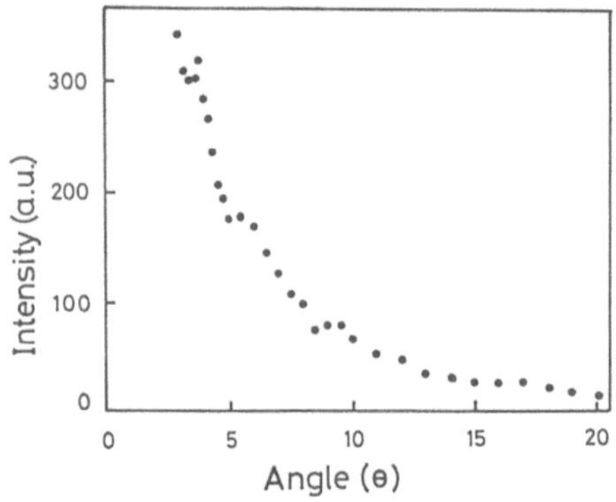

Fig. 4. Light scattering profile of a PSA/P2CS submitted to a shallow jump and irradiated in 5min.

of instability originating from the photo-cross-link reactions taking place in irradiated blends. Further experiments are currently in progress to elucidate these light-induced instabilities.

Acknowledgements

We are deeply indebted to Dr. M. Okada (Tokyo Institute of Technology) for the light scattering instrument and suggestions. Helpful discussions with Drs. S. Sakurai and M. Shibayama are gratefully acknowledged.

References

1) G. Nicolis and I. Prigogine, "Self-Organization Process in Non-Equilibrium Systems", Wiley-Interscience, New York (1977).

2) "Nonlinear Phenomena in Materials Science", L. Kubin and G. Martin Eds., Trans Tech Publication, Switzerland (1988).

3) Q. Tran-Cong, T. Chang and C. C. Han Polymer 29, 2261 (1988).

4) Q. Tran-Cong, T. Nagaki, T. Nakagawa, O. Yano and T. Soen Macromolecules 22, 2720 (1989).

Part III

Liquid Crystals

Ordered Phases in Colloidal Suspensions of Tobacco Mosaic Virus

R.B. Meyer

The Martin Fisher School of Physics, Brandeis University,
Waltham, MA 02254, USA

Abstract. Suspensions of Tobacco Mosaic Virus in water have a remarkable phase diagram exhibiting a number of ordered states. To some extent the ordering can be understood in terms of hard particle interactions, which makes the variety of ordered states all the more remarkable. This article discusses our current knowledge of the Tobacco Mosaic Virus system, some ideas about the types of order observed, and the possibilities for future research.

1. Introduction

Colloidal suspensions of Tobacco Mosaic Virus in water are unique in several ways. Basically, they are the only readily reproducible systems of identical rod–like colloidal particles. As a consequence of the rod–like particle shape, a remarkable variety of ordered phases of this system occur when the particle concentration is sufficient for the particles to interact strongly. These include various liquid crystal and crystalline phases. Historically Tobacco Mosaic Virus has played a special role in the development of the theory of liquid crystal ordering, and it continues to be important in that respect.

In recent years, much has been learned about the ordered states of Tobacco Mosaic Virus. It is also clear that much remains to be learned. This article will attempt to introduce the subject, combining basic ideas about ordering and experimentally determined facts about Tobacco Mosaic Virus as a colloidal particle. It will then discuss the recent development of ideas and experiments concerning ordering of Tobacco Mosaic Virus, and finally indicate a number of open questions. This paper concentrates on work done at Brandeis University.

2. Tobacco Mosaic Virus as a Colloidal Particle

Tobacco Mosaic Virus (TMV) has a simple structure, compared to many other viruses. It consists of 2130 identical protein molecules arranged in a simple helix to form a cylindrical rod 180Å in diameter and 3000Å long, with a hollow core 40Å in diameter. Coiled in a groove between the protein molecules in the helix is an RNA molecule whose length determines the length of the virus particle [1].

Springer Proceedings in Physics, Vol. 52 Dynamics and Patterns in Complex Fluids
Editors: A. Onuki · K. Kawasaki

As evidenced by freeze fracture electron micrographs [2], these particles are rigid rods, with a bending elastic modulus that is large enough that thermally excited curvature of the rod is negligible [3].

Each protein in the virus has ionic groups on its surface that tend to be dissociated in water. The isoelectric point is pH 3.3, so under neutral or slightly basic conditions each TMV particle has a potentially large negative charge [4]. Due to the highly nonlinear nature of the Poisson–Boltzman equation, much of this charge is cancelled by closely associated counterions. The resulting screened Coulomb repulsion between virus particles is characterized by a charge of a few thousand electrons per particle. In solutions of low enough ionic strength this leads to a stable colloidal suspension [5].

However, these particles are less than ideal in several ways. The proteins are not strongly bonded together, so the particles can be broken. They can also disassemble due to electrostatic repulsion among the proteins, if the ionic strength of the surrounding solution is too low. On the other hand, at too high ionic strength, the particles can stick together end to end, with the helical protein stack in one virus simply continuing in the next one. This event is irreversible, so if it occurs at any time in the extraction, purification, or handling of a sample of the virus, it permanently changes the sample. One can prepare intentionally aggregated samples, to study very long particles, but in the experiments described here we studied preparations that were as close to monodisperse in undamaged and unaggregated virus as possible [6].

3. Ordering of Rod-Like Particles

In 1949, stimulated by the observation of nematic ordering in colloidal suspensions of TMV, Onsager proposed a fundamental model for the parallel ordering of rod–like particles [7]. He argued that excluded volume effects, rather than long range forces, dominated the ordering, so that TMV particles should be thought of primarily as hard rods. Nematic ordering must therefore maximize the entropy of the system at high enough concentration of rods. The volume Δv excluded for the center of one rod by the presence of another is

$$\Delta v \cong 2d\, L^2 \mid \sin\theta \mid \;,$$

in which L is the length and d the diameter of the rods ($L >> d$) and θ the angle between their axes. We see that this leading term in Δv is minimized by making the rods parallel. Onsager used a second virial expansion of the free energy F per particle:

$$F/NkT = \mu_0(T) - 1 + \ln c + \sigma(f) + bc\rho(f) \;,$$

$$\sigma(f) = \int f(\mathbf{a}) \ln[4\pi f(\mathbf{a})]\, da \ ,$$

$$\rho(f) = \frac{4}{\pi} \int K(\mathbf{a}, \mathbf{a}') f(\mathbf{a}') f(\mathbf{a})\, da\, da' \ .$$

Here $\mu_0(T)$ is the chemical potential of a particle in the solvent at temperature T, $c = N/V$ is the particle number density, and $b = \pi L^2 d/4$. Unit vectors $\mathbf{a}$ and $\mathbf{a}'$ give the orientations of the rods. K is proportional to the excluded volume Δv.

Onsager showed that above a critical concentration, nematic ordering maximizes the entropy of the system. This is fundamentally a competition between two components of the entropy, one, σ, describing the loss of orientational entropy with the appearance of long range nematic order, and the other, $bc\rho$, describing the increase of configurational entropy for the centers of mass of the particles, due to the reduction of excluded volume via parallel ordering. Note that the first term is a single particle entropy, while the second is an approximation to a collective effect; the volume made available by parallel ordering is shared by all the particles. Therefore at high concentrations the second term must dominate, and ordering will maximize the entropy.

It can be argued that this trend is the driving principle for the appearance of long range order in any hard particle system. The appearance of the long range order reduces a single particle contribution to the entropy, while it makes available free volume that increases the entropy of the system much more strongly at high concentration. The most extreme case of this argument accounts for the crystallization of hard spheres. In that case, one tends to say that crystallization is the most efficient packing, which is equivalent to saying it makes the most volume available for rearrangements of all the particles. We will see applications of this argument to phases of TMV that are more ordered than the nematic, but less than fully crystalline.

4. Nematic Ordering

For all the insight of Onsager's original article, it did have its limitations, and surprisingly enough until recently there was not much follow–up on his work. Recently, there has been much more research, both theoretical [8]–[21] and experimental. Although Onsager's simple mathematical model embodies the essential physics of the nematic ordering of hard rods, it is applicable only to rather long rods, so that all the interesting effects take place at very low concentration. Moreover, TMV is not simply a hard rod, but is a charge stabilized colloidal particle, with a finite range screened coulomb repulsion between rods. Onsager was aware of this, and addressed the effects of charge in terms of an effective particle diameter that depended on surface charge and ionic strength of the solution.

In recent years, a number of steps have been made to extend the Onsager model to the study of elastic properties of the nematic phase [15], to improve the quantitative predictions of the model for the ordering of short rods interacting via the screened coulomb interaction [10][11], and to apply the model to the study of pretransition ordering of the rods in the isotropic phase [8]. Equally important, with improvements in the preparation of high quality TMV samples, a number of experiments have been performed to test the Onsager theory and its modern extensions, opening the way to deeper understanding of the nematic ordering in the TMV system.

Two independent groups found an algorithm for the accurate solution of Onsager's original integral equation by an iterative method [16][17]. Using this calculational method, which is accurate at all particle concentrations, Straley's method of calculating the nematic elastic moduli [15] was carried out accurately up to high concentrations [17]. Besides obtaining numerical predictions of the elastic moduli, an important qualitative trend became clear; the bend modulus diverges at high concentration, while the splay and twist moduli remain small. This has to do with two factors. First, at high concentration, the degree of parallel ordering of the rods becomes very large, so that the mutual interference effects of the rods are minimized. For splay and twist distortions, in which the director field can still be described as straight lines, this means that the extra interparticle interferences induced by the distortions become smaller as order becomes perfect, so the splay and twist elastic moduli do not grow with increasing concentration. For the bend distortion, however, the director field is fundamentally bent, so that even at perfect parallel ordering of the rods, the bend creates significant new interference effects. Moreover, because the excluded volume function is singular in the interparticle angle θ, at $\theta = 0$, for perfect parallel ordering, the bend modulus would become infinite. These facts combine to make the bend modulus become very large at high concentration.

Another major step in adapting the Onsager model to real systems was the development of a functional scaling method developed by Sin-Doo Lee [18] for interpolating between the Onsager free energy for very long hard spherocylinders and the Carnahan–Starling expression for the free energy of hard spheres, *i.e.*, spherocylinders of zero cylindrical length. This new free energy expression, combined with the iterative solution of the resulting generalization of the Onsager integral equation, produces a phase diagram in good agreement with the few available computer simulations of short hard rod systems. The analogous calculation for ellipsoidal particles also produced excellent agreement with simulations [19].

It is remarkable that until recently, detailed experimental tests of the Onsager theory have not been performed. What was needed were high quality TMV samples and a method of measuring the orientational distribution function of the rods. With the development of excellent sample preparation methods in recent years [6] it remained to develop a measurement of the distribution function. This was done using x-ray diffraction [22].

The method is based on several properties of the TMV system. First, the single particle fourier transform is a thin flat disc in reciprocal space. Using Cu K_α radiation, the thickness of the disc is much less than the resolution of most small angle scattering set-ups. Second, the mean lateral interparticle spacing in the TMV nematic phase is significantly larger than the diameter of a single particle, so that at large enough scattering angles, there is still measurable intensity from the single particle diffraction, but interparticle interference effects have disappeared.

The experiment is performed with the x-ray beam incident on a nematic TMV sample aligned in a magnetic field with the nematic alignment axis, or director, perpendicular to the x-ray beam. The scattering is recorded on film oriented normal to the beam, about 30 cm from the sample. There is an intense streak of scattering along the equatorial direction, normal to the director. The smearing of this streak above and below the equator is due to the angular distribution of the TMV particles. At small angles, the smearing is a function of interparticle interference effects, but at large angles (above about 1°) it is simply a representation of the single particle angular distribution function. It is simple to calculate the intensity distribution function on the film from the particle distribution [22], and the inverse transformation has also been derived [23].

The measured angular distribution function is remarkably close to a gaussian peaked on the director. This is very close to the form predicted by the Onsager theory [7]. Using Lee's modification of the Onsager theory [24], one can make a detailed comparison of the measured angular distribution function, as a function of concentration, with theory. An important parameter for this comparison is the effective particle volume, due to the screened Coulomb interaction. By fitting the measured particle concentration at the nematic–isotropic coexistence to the theoretical volume fraction at that point, one can derive an effective particle volume. For the case in which this was done, this amounted to increasing the diameter of the TMV particle by about 35Å, which is consistent with the estimated Debye screening length for the solution. The measured nematic order parameter (the second Legendre polynomial moment of the angular distribution function) was also within a few percent of the theoretical value through the nematic range. These results were the first detailed quantitative experimental confirmation of Onsager's ideas.

A second experiment which further tests the Onsager model, and Lee's extension of it, is the measurement of the nematic–isotropic phase transition as a function of the ionic strength of the solution [25]. Since this is a first order phase transition, there is a finite range of mean particle concentration over which the nematic and isotropic phases, with different concentrations c_n and c_i, are in equilibrium. This is a consequence of the fact that a sealed colloidal sample is effectively a gas of particles at fixed volume, rather than fixed pressure. As one increases the ionic strength, the effective particle volume becomes smaller, so the mean phase transition concentration increases. The Lee calculation predicts this accurately. However, the measured difference between c_n and c_i is found to be quite a bit larger than that calculated by Lee.

Further experiments that have examined the interactions that lead to nematic ordering of TMV focus on the correlations between rods that develop in the isotropic phase at concentrations below that for nematic ordering [25]. One sees evidence of this pretransitional ordering by measuring the magnetically induced birefringence of the sample as a function of TMV concentration. Further information is gained by again varying the ionic strength, and thus the range of the electrostatic repulsion between rods. Again, the data compare very well with calculations based on an extension of Lee's model. Experiments on the temperature dependence of the magnetic birefringence [26][27] have raised questions about the purely repulsive nature of the interparticle interaction in the Onsager picture [8][28]. It appears that this question may be resolved by considering the nature of movements of the diamagnetically anisotropic benzene rings in the TMV structure, which result in a temperature dependent anisotropy to the magnetic susceptibility for the particle [29].

5. Smectic Ordering

The first strong evidence for smectic ordering of hard rods came from computer simulations of the system of parallel spherocylinders [30]. Remarkably, the simulations gave clear evidence of a second order phase change from nematic to smectic A ordering at a particle density well below that for crystallization. Previous to these results, there was a wide spread belief that smectic ordering was a consequence of the details of the pairwise molecular interaction, in which rigid "core" parts and flexible "tail" parts of molecules tended to segregate into layers, for energetic or entropic reasons. That simple rigid cylinders should spontaneously form a phase consisting of a stack of two dimensional fluid layers was a surprise.

Recalling the Onsager argument for nematic ordering, can one devise a separation of the entropy into competing terms such that smectic ordering maximizes the entropy at high concentration? To see that this is possible, one has to consider how the unoccupied volume in the nematic and smectic phases is distributed.

For parallel cylinders in the nematic phase, it becomes evident that there is volume adjacent to the ends of each cylinder that is unlikely to be occupied by other cylinders. This is because in the nematic phase, cylinders next to one another tend to overlap lengthwise by random amounts, and any overlap prevents them from moving into an end-to-end configuration. In the smectic phase, however, neighboring molecules tend to be in the same layer, and therefore to overlap one another lengthwise by a large amount. Their ends tend to be aligned with one another, with a relatively small gap between them and the ends of molecules in the next layer. The isolated volume elements at the ends of molecules in the nematic phase have been eliminated; those volume elements were available only to single molecules, while in the smectic phase that volume has been redistributed so that it is available for configurations of all the molecules. Therefore at high density the smectic phase is favored.

Another way to see this clearly is to look at the highly ordered smectic phase, with well defined layers. If a molecule partially projects into a neighboring layer, it occupies area in two layers instead of one, and creates void spaces adjacent to its ends that are available almost exclusively to itself.

This argument suggests a model for calculation of the entropy [20], in which one defines a two dimensional packing density in a layer by counting all the molecules that occupy area in the layer. This includes all cylinders whose centers lie one cylinder length above or below the center of the layer. In the well ordered smectic phase, with layers a little more than a cylinder length in thickness, this counts cylinders only in one layer, while in the nematic phase, arbitrarily choosing a layer, this tends to include many more molecules. At high three-dimensional particle densities, this means that the nematic "layer" is very tightly packed with little entropy, while the smectic layers contain fewer molecules and therefore have many more internal configurations. This model calculation produces an expression for the entropy that spans the nematic and smectic phases continuously, although it probably does not describe the nematic phase very well. It predicts a second order nematic–smectic phase transition and a smectic layer thickness in good agreement with the simulations for parallel cylinders, but the transition occurs at an unrealistically low particle density.

Another calculation based on the same idea, that creating a periodic ordered state uses volume more effectively at high densities, is carried out by literally partitioning space into layers with hard boundaries in the smectic state [21]. By varying the layer thickness at total fixed volume, and using an accurate model for calculating both the entropy of the nematic phase and the entropy of the same number of molecules in the two dimensional layer state, one finds an accurate estimate of the nematic–smectic transition density, although the model is intrinsically incapable of producing a second order transition.

The shape of the hard particles is important to the appearance of smectic ordering [20]. Rod-like ellipsoids do not produce a smectic phase. Because the ends of ellipsoids are tapered, the argument based on the creation of volume-wasting voids is weakened, so it is easier for particles to penetrate neighboring smectic layers, without immediately occupying a large area. In fact, as an ellipsoid moves partially out of one layer, unlike a cylinder, it also occupies less area in that layer. These subtle changes are enough so that nematic ordering persists to high densities, above which crystallization appears.

Between these two calculations based on a clear physical argument, neither one of which captures all the quantitative elements of the nematic–smectic transition, one is reasonably convinced of the argument. Less intuitive calculations based on the density functional formalism predict well the properties of the transition seen in the simulations [31]. It should be noted that in simulations in which the condition of strictly parallel rods was relaxed, the nematic–smectic phase transition still occurred, at a slightly higher density. With this rather successful picture for hard rod smectic ordering, it remained to be seen if smectic ordering occurred in a real physical system resembling hard rods.

For years, layered structures, usually presumed or observed to be crystalline, have been reported in TMV samples [32]. Now, definitive evidence of smectic ordering has now been found in TMV, under conditions in which the particles interact as reasonably hard rods [33]. The smectic ordering was discovered by examining samples of different compositions. Because the TMV particle length places the smectic layer thickness in the visible optical range, the most obvious characteristic of the smectic phase is the brilliant iridescence of the sample due to Bragg scattering from the layers. Moreover, one sees that the smectic domain structure is quite unlike that of crystalline ordering, with curved layers and fan–like patterns, and the characteristic focal conic defect lines of the classic smectic phase. The definitive evidence completing the confirmation of the smectic phase was the observation of two dimensional liquid ordering of the virus particles in the layers, measured by x-ray diffraction.

With little knowledge yet of the detailed properties of the TMV smectic phase, it is premature to make too close a comparison with the hard rod ordering picture, although it is tempting to do so, since the systems have traditionally been so closely associated with one another. As with the nematic–isotropic transition, if one looks at the nematic–smectic transition in TMV, and demands that the effective particle volume be adjusted so that the volume fraction at the transition matches that of the hard rod transition, the particle volume necessary is again slightly larger than the bare particle volume. This is again reasonably consistent with the effective volume expected from the screened coulomb interaction. The Debye screening length is about 10% of the particle diameter, so the particles interact as if they are effectively cylindrical in shape. It is notable that at lower ionic strength the smectic phase is not seen, and there is a direct transition from nematic to crystalline order. At lower ionic strength, the particles interact with a more ellipsoidal effective shape. In computer simulations, the system of hard ellipsoids also does not exhibit a smectic phase. To the extent that one can model the screened coulomb interaction with an effective hard particle shape, these results are consistent, but that may be a coincidence.

There are some clear differences between the TMV smectic phase and the hard rod smectic. The current observation is that the TMV smectic phase exists in a very narrow concentration range between the nematic and crystal phases. Moreover, it looks like the nematic–smectic transition is first order, although there are clear pretransitional smectic fluctuations in the nematic phase, so the transition may be only weakly first order. The smectic layer spacing in TMV is only 12% larger than the virus length, while in the hard rod system it is found to be 27% larger than the rod length. It may be that the fundamental physics that leads to the smectic ordering in TMV is based on its resemblance to a hard rod, and that the residual effects of the soft coulomb repulsion account for the differences between the smectic ordering in the two systems. However, the differences may be more fundamental. More experiments are needed to explore these questions further.

6. Columnar Ordering

The computer simulations that produced smectic ordering of parallel hard rods also produced columnar ordering at higher densities, for long spherocylinders. The columnar state, a two dimensional hexagonal crystal lattice of columns in which there is dense one dimensional liquid order within each column, is the other major state possible, besides smectic, that is partially crystalline. At very high densities, the smectic layers would crystallize internally, and probably immediately interact to form a three dimensional crystal. In the columnar phase, density comparable to that of a crystal is obtained with some residual entropy remaining in the liquid order in each column and the lack of correlations between the liquids in neighboring columns. Again here, as for the smectic, the detailed shape of the particles is crucial. Cylindrical particles in neighboring columns can slide past one another without interference, while ellipsoidal particles cannot.

The same model calculations which restricted particles to layers to justify existence of the smectic phase also produced the columnar phase by restricting particles to tubes, representing the columns of the columnar phase [21]. Once again, trading the entropy lost in creating long range order for extra free volume to facilitate more fluid particle configurations maximizes the entropy at high enough density.

Does columnar order exist in TMV? The evidence from our experiments is not yet definitive. There appears to be a new phase in very low ionic strength samples at TMV concentrations above that of the crystal phase! X-ray experiments show that there is hexagonal ordering of the virus rods in this phase, and the optical texture observed in the polarizing microscope is fibrous, quite different from that of the compact crystalline grains seen at lower concentration. The phase is weakly iridescent, but we have not yet obtained high quality single domain samples for the necessary quantitative analysis of the optical diffraction, to distinguish columnar ordering from some kind of plastically deformed crystalline texture. The samples are quite viscous, as would be expected for columnar order or for some modification of crystalline order. It is not clear that the samples are in an equilibrium state.

7. Open Questions

The phase diagram of TMV as a function of particle concentration, ionic strength, and pH, which controls the surface charge on the virus, is still quite incomplete. Even where phases have been identified, we are just beginning to measure their properties accurately. What is very exciting is that with each new piece of information, one can test ideas about the molecular interactions, the fundamental nature of the ordering, and the extent to which our current picture is right. This is because the system is simple enough both conceptually and in many practical aspects so that one can develop testable theories and do clear experiments. An example of this is the development of the hydrodynamic theory of the fluctuation modes of the TMV colloidal smectic phase. Because the microscopic picture of colloidal particles arranged in layers, imbedded in a fluid solvent is a simple one, one can develop clear predictions for the nature of the low energy, long wavelength fluctuation modes of this system [34]. Experiments to test the hydrodynamics are in progress [35].

In understanding the nature of the ordering of TMV, the detailed role of electrostatic interactions is not yet completely clear. Equally important is the role of polydispersity due to the inevitable presence of at least small fractions of broken or aggregated particles. This can have profound effects on the phase diagram. For instance, it can easily confuse the question of the first or second order nature of the nematic–smectic phase transition. Polydispersity clearly disfavors layered structures, which are most compatible with particles identical in length. This argument also indicates that the question of the existence of columnar order in TMV is a delicate one, since columnar order is quite compatible with a distribution of particle lengths, and the columnar phase at high density is competing with layered phases.

I have not discussed the crystalline phase here, but it offers opportunities for fascinating experiments. With the discovery of the smectic A phase, we have also seen beautiful well oriented barrel–shaped single crystals of higher TMV concentration suspended in alignment with the smectic layers [36]. X-ray and light scattering experiments on these samples are under way.

It is clear that we have much more to learn from Tobacco Mosaic Virus. It is a unique system, but it serves as a model and testing ground for much more generally applicable ideas. The deeper understanding of TMV has implications for spatially ordered systems in general, and clearly for all liquid crystalline systems. One day we may have other reliably reproducible rod-like colloidal particles to study, and compare to TMV, which will only make this simple virus that much more interesting.

8. Acknowledgements

The work discussed here that was done at Brandeis, or closely connected with it, has been the result of collaboration by many people. Don Caspar and the general participation of the structural biology laboratory in the Rosensteil Center, especially Marguerite Cahoon and Walter Phillips, have been central to this research program. Faculty colleagues, postdoctoral research associates, and graduate students working with me, including Alan Hurd, Rudolf Oldenbourg, Frank Lonberg, Sin-Doo Lee, Seth Fraden, Xin Wen, and Xiaolei Ao have all made important contributions. Both Rudolf and Seth benefitted greatly from their association with Georg Maret in Grenoble, where part of the work was done. Our interactions with Judy Herzfeld, Mark Taylor, and Reinhard Hentschke, in the chemistry department have been very stimulating as well.

This research was supported in part by the National Science Foundation through grant DMR-8803582, and by the Martin Fisher School of Physics, Brandeis University.

References

[1] D.L.D. Caspar, Adv. Protein Chem. **18**, 37 (1963).

[2] J.A.N. Zazodzinski, Phys. Rev. Lett. **56**, 636 (1986).

[3] E. Loh, Biopolymers **18**, 2549 (1979); **18**, 2569 (1979).

[4] R.B. Scheele and M.A. Lauffer, Biochemistry **6**, 3076 (1967).

[5] V.A. Parsegian and S.L. Brenner, Nature (London) **259**, 632 (1976).

[6] Samples were prepared by a procedure based on that of H. Boedtker and N.S. Simmons, J. Am. Chem. Soc. **80**, 2550 (1958).

[7] L. Onsager, Ann. N.Y. Acad. Sci. **51**, 627 (1949).

[8] J.P. Straley, Mol. Cryst. Liq. Cryst. **22**, 333 (1973).

[9] J.P. Straley, Mol. Cryst. Liq. Cryst. **24**, 7 (1973).

[10] Th. Odijk, Macromolecules **19**, 2313 (1986).

[11] A. Stroobants, H.N.W. Lekkerkerker, and Th. Odijk, Macromolecules **19**, 2232 (1986).

[12] M.A. Cotter, Phys. Rev. A. **10**, 625 (1974).
[13] G. Lasher, J. Chem. Phys. **53**. 4141 (1970).
[14] H.N.W. Lekkerkerker, Ph. Coulon, R. Van Der Haegen, and R. Deblieck, J. Chem. Phys. **80**, 3427 (1984).
[15] J.P. Straley, Phys. Rev. A. **8**, 2181 (1973).
[16] J. Herzfeld, A.E. Berger, and J.W. Wingate, Macromolecules **17**, 1718 (1984).
[17] S.D. Lee and R.B. Meyer, J. Chem. Phys. **84**, 6 (1986).
[18] S.D. Lee, J. Chem. Phys. **87**, 8 (1987).
[19] S.D. Lee, J. Chem. Phys. **89**, 7036 (1988).
[20] X. Wen and R.B. Meyer, Phys. Rev. Lett. **59**, 1325 (1987).
[21] M.P. Taylor, R. Hentschke, and J. Herzfeld, Phys. Rev. Lett. **62**, 800 (1989).
[22] R. Oldenbourg, X. Wen, R.B. Meyer, and D.L.D. Caspar, Phys. Rev. Lett. **61**, 1851 (1988).
[23] M. Deutsch, Phys. Rev. Lett. **64** 697 (1990).
[24] S.D. Lee, J. Chem. Phys. **87**, 4972 (1987).
[25] S. Fraden, G. Maret, D.L.D. Caspar, and R.B. Meyer, Phys. Rev. Lett. **63**, 2068 (1989).
[26] P. Photinos, C. Rosenblatt, T.M. Schuster, and A. Saupe, J. Chem. Phys. **87**, 6740 (1987).
[27] S. Fraden, to be published.
[28] P. Photinos, and A. Saupe, Mol. Cryst. and Liq. Cryst. **123**, 217 (1985).
[29] S. Fraden, private communication.
[30] A. Stroobants, H.N.W. Lekkerkerker, and D. Frenkel, Phys. Rev. Lett. **57** 1452 (1986);Phys. Rev. A. **36**, 2929 (1987); D. Frenkel, H.N.W. Lekkerkerker, and A. Stroobants, Nature (London) **332**, 822 (1988); D. Frenkel, J. Phys. Chem. **92**, 3280 (1988).
[31] B.M. Mulder, Phys. Rev. A **35**, 3095 (1987); A. Poniewierski and R. Holyst, Phys. Rev. Lett. **61**, 2461 (1988); A.M. Samoza and P. Tarazona, Phys. Rev. Lett. **61**. 2566 (1988).
[32] G. Oster, J. Gen. Physiol. **33**, 445 (1950); U. Kreibig and C. Wetter, Z. Naturforsch, **35c**, 750 (1980); S. Fraden, W.C. Phillips, and D.L.D. Caspar, Biophysical J. **37**, 97a (1982); S. Fraden, A.J. Hurd, M. Cahoon, D.L.D. Caspar, and R.B. Meyer, J. Phys. (Paris), Colloq. C3, **46**, C3-85 (1985).
[33] X. Wen, R.B. Meyer, and D.L.D. Caspar, Phys. Rev. Lett. **63**, 25 (1989)
[34] X. Wen and R.B. Meyer, J. Phys. France **50**, 3043 (1989).
[35] X. Ao and X. Wen, in preparation.
[36] X. Wen, in preparation.

A Model of Electro-rheological Fluid

J. Takimoto

Scientific Research Laboratory, Central Engineering Laboratories,
Nissan Motor Co., Ltd., 1 Natsushima-cho, Yokosuka 237, Japan

Abstract. A simple model of electro-rheological fluid is presented, and some of its properties are studied by dimensional analysis and numerical simulation.

1. Introduction

Electro-rheological fluid (ERF) is a dispersion of highly polarizable particles in insulating liquid, and characterized by dramatic change of its rheological properties by external electric field (see [1] and references therein). To be more precise, consider the shear stress τ (xz-component of stress tensor) of the ERF under the shear flow $u_x = \dot{\gamma} z$. With no electric field it is an ordinary Newtonian fluid (τ is proportional to the shear rate $\dot{\gamma}$). If, on the other hand, the electric field E_ϵ is applied in the z-direction, the ERF behaves as a Bingham fluid and τ depends on $\dot{\gamma}$ as [1]

$$\tau = \tau_y + \eta\dot{\gamma}. \tag{1}$$

with the yield stress τ_y proportional to E_ϵ^2. By virtue of this property, the ERF has many possible applications, including active dampers, robotic controls, etc.

2. Model

In order to clarify the basic mechanism of ERF, we consider the following simple model: Dispersed particles are metallic spheres of radius a, and dispersing medium is insulating liquid with dielectric constant ϵ_0 and viscosity η_0. Mass densities ρ of the particles and the liquid are taken to be equal. Typical values of the parameters are $a \sim 1\mu$m, $\epsilon_0 \sim 1$, $\eta_0 \sim 0.1$ poise, $\rho \sim 1$ g/cm^3, $E_\epsilon \sim 1$ kV/mm and $\dot{\gamma} \sim 10^3$/sec. Besides the volume fraction ϕ of the dispersed phase, we can construct three independent dimensionless parameters:

$$P = \frac{k_B T}{a^3 \epsilon_0 E_\epsilon^2} \sim 10^{-4}, \qquad Q = \frac{\rho a^2 \epsilon_0 E_\epsilon^2}{\eta_0^2} \sim 10^{-3}, \qquad R = \frac{\rho a^2 \dot{\gamma}}{\eta_0} \sim 10^{-4}. \tag{2}$$

$P \ll 1$ means that the effect of thermal motion (Brownian motion) of the particles can be neglected. Both Q and R can be regarded as Reynolds numbers, and smallness of them indicates that the inertial effect can be neglected. In other words, velocity of each particle is so determined that the Stokes resistance is balanced with the electrostatic force.

Springer Proceedings in Physics, Vol. 52 Dynamics and Patterns in Complex Fluids
Editors: A. Onuki · K. Kawasaki

Using a dimensionless function $f(\phi, Q, R)$, the shear stress τ can be expressed as $\tau = (\eta_0^2/\rho a^2) f(\phi, Q, R)$. If we assume that f can be expanded in powers of Q and R as $f(\phi, Q, R) \approx f_1(\phi)Q + f_2(\phi)R$, we can reproduce the experimental behavior (1) with

$$\tau_y = \epsilon_0 E_e^2 f_1(\phi), \qquad \eta = \eta_0 f_2(\phi). \tag{3}$$

Note that τ is independent of a and ρ in this approximation. To estimate the order of magnitude of τ_y, let us assume $f_1 \sim 1$. Then we get $\tau_y \sim 10^3\mathrm{dyn/cm^2}$ for $E_e \sim 1\mathrm{kV/mm}$. Experimental values are around $10^2 \sim 10^4$ dyn/cm^2.

3. Effect of Clusters

When $\dot{\gamma} = 0$, the electric field causes the formation of chain-like clusters. We are numerically studying this effect by extending the method for the two-particle case [2] as follows: interactions among particles are approximated by sums of two-body forces. Moreover, we restrict center of each particle to be on a single plane. Results of our simulation (Fig. 1) indicate that the formation of clusters proceeds in two stages: (i) "short" clusters (2 or 3 particles) form in the time $\sim 30\eta_0/(\phi^{5/3}\epsilon_0 E_e^2)$, and (ii) longer clusters form by aggregation of short clusters on a much longer time scale.

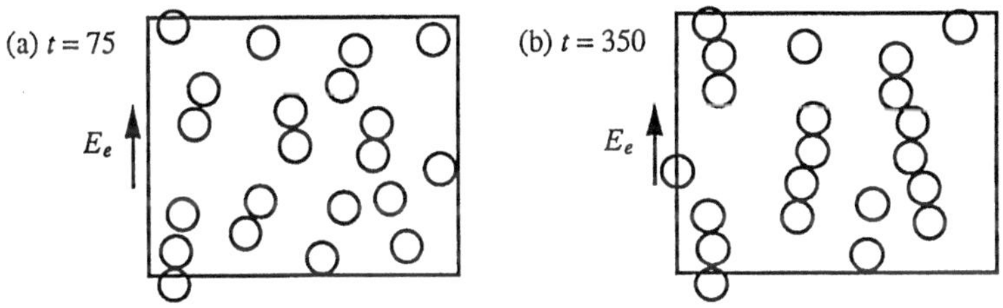

Fig. 1. Results of simulation of cluster formation process ($\dot{\gamma} = 0$). Surface of each particle is coated by insulating layer of thickness $0.2a$. Unit of time: $t_0 = 6\pi\eta_0/\epsilon_0 E_e^2$. Periodic boundary condition is used.

What happens to these clusters when shear flow exists ? If ϕ is so small that interactions between clusters can be neglected, it can be argued that clusters of length up to $n_c \propto (\epsilon_0 E_e^2/\eta_0\dot{\gamma})^{1/2}$ survive, and $f_1(\phi) \propto \phi$. If, on the other hand, ϕ is not small, our preliminary simulations indicate that collisions among clusters destroy longer clusters and only short clusters can exist. In order to estimate the shear stress in this case, simulations with a larger number of particles are now in progress.

References

[1] H. Block and J. P. Kelly, *J. Phys. D: Appl. Phys.* **21**, 1661 (1988)
[2] P. A. Arp and S. G. Mason, *Colloid Polymer Sci.* **255**, 566 (1977)

The Dynamical Structure Factor of Rodlike Polymers in the Isotropic Phase

T. Maeda

Mitsubishi Kasei Institute of Life Sciences,
11 Minamiooya, Machida-shi, Tokyo 194, Japan

The self-diffusivities of rodlike polymers in the semidilute regime are estimated as [1]

$$D_{\parallel} \simeq D_{\parallel 0}, \qquad D_{\perp} \simeq 0, \qquad D_{\mathrm{r}} \simeq D_{\mathrm{r}0}/[1+\beta(\nu L^3)^2], \tag{1}$$

where $D_{\parallel}$ and $D_{\perp}$ are respectively the lengthways and sideways translational diffusion coefficients of a rod, D_{r} is the rotational one, subscript 0 of $D_{\parallel 0}$, and $D_{\mathrm{r}0}$ stands for zero polymer concentration, ν is the number concentration, and L the rod length. Here β is an empirical constant of the order of 10^{-3}.

If we use the dynamic light scattering (DLS) technique to study the diffusive motion of a rod in the less-dilute regime, we should note that the measured apparent diffusion coefficient $\Gamma_{\mathbf{k}}/k^2$ is an average of the mutual diffusion coefficients ${}^*D_i$ for $i =$ "$\parallel$", "$\perp$", and "r", where $\Gamma_{\mathbf{k}}$ is the initial decay rate, and $\mathbf{k}$ the scattering vector. The mutual diffusion coefficient ${}^*D_i$ is related to the self-diffusivity as ${}^*D_i = D_i A(\nu)$, where $A(\nu)$ expresses the effect of the intermolecular interaction. When the polymer concentration is very low, both diffusion coefficients should agree with each other, i.e., $A(\nu) \to 1$ for $\nu \to 0$. Using the polarized DLS technique, Russo et al. [2] studied the concentration dependence of the apparent diffusion coefficient $\Gamma_{\mathbf{k}}/k^2$ of poly(γ-benzyl L-glutamate) (PBLG). They concluded that the A term was sufficiently strong to cause an *increase* in the measured $\Gamma_{\mathbf{k}}/k^2$ even with the decreased D_i values in the semidilute regime, but no theory for the A term was available until recently.

It was Doi, Shimada and Okano [3] who opened the path to the formulation of the A term. Using the mean-field approximation for the hard-rod interaction potential, they derived an expression for the dynamical structure factor $g(\mathbf{k},t)$ of rodlike polymers for the isotropic scattering, where t is the delay time. Unfortunately, it being difficult to evaluate their $g(\mathbf{k},t)$, they gave exact expressions only for two limiting cases, i.e., $\mathbf{k} \to 0$ and $t \to 0$.

Looking for a more tractable form of the dynamical structure factor, we rewrote their $g(\mathbf{k},t)$ using vectors and matrices [4]. Following the same line of argument, we also derived the dynamical structure factor $g_{\mathrm{VH}}(\mathbf{k},t)$ for the VH-depolarized scattering [5]. They are given by the following equations:

$$g(\mathbf{k},t) = L\mathbf{s}^{\mathrm{T}} \exp[-\boldsymbol{\Omega} t]\boldsymbol{\Phi}^{-1}\mathbf{s}, \tag{2}$$

$$g_{\mathrm{VH}}(\mathbf{k},t) = L\mathbf{d}^{\mathrm{T}} \exp[-\boldsymbol{\Omega} t]\boldsymbol{\Phi}^{-1}\mathbf{d}. \tag{3}$$

Here $\mathbf{s}$ and $\mathbf{d}$ are column vectors that stand for the electric field of the isotropically and VH-depolarized scattered light, respectively, and superscript T of $\mathbf{s}^{\mathrm{T}}$ and $\mathbf{d}^{\mathrm{T}}$ denotes the transposition. $\boldsymbol{\Omega}$ is a matrix that expresses the translational-rotational diffusion of a rod under the influence of other polymers: $\boldsymbol{\Omega} = \boldsymbol{\Theta}\boldsymbol{\Phi}$, where matrix $\boldsymbol{\Theta}$ expresses

Springer Proceedings in Physics, Vol. 52 Dynamics and Patterns in Complex Fluids
Editors: A. Onuki · K. Kawasaki

the diffusion of a rod free from the hard-rod interaction with other polymers, and matrix $\mathbf{\Phi}$ stands for the mean-field interaction. Matrix $\mathbf{\Phi}$, of course, reduces to the identity matrix $\mathbf{E}$ at zero polymer concentration. Though eq 2 holds only for the isotropic scattering, it is almost always applicable to the polarized scattering, because the polarizability anisotropy of rodlike polymers is usually small compared with the average molecular polarizability.

We have the following two equations for the excluded volume effect term A.

$$A(\nu) = (\mathrm{s}^{\mathrm{T}}\mathrm{s})/(\mathrm{s}^{\mathrm{T}}\mathbf{\Phi}^{-1}\mathrm{s}), \tag{4}$$

$$A_{\mathrm{VH}}(\nu) = (\mathbf{d}^{\mathrm{T}}\mathbf{d})/(\mathbf{d}^{\mathrm{T}}\mathbf{\Phi}^{-1}\mathbf{d}), \tag{5}$$

where $A(\nu)$ is for the isotropic scattering and $A_{\mathrm{VH}}(\nu)$ for the depolarized one. Since $\mathbf{\Phi} = \mathbf{E}$ for $\nu = 0$, it is evident that $A(0) = A_{\mathrm{VH}}(0) = 1$, as it should be. $A(\nu)$ increases with increasing ν, but $A_{\mathrm{VH}}(\nu)$ decreases: The excluded volume effect causes an *increase* in the initial decay rate $\Gamma_{\mathbf{k}}$ of $g(\mathbf{k},t)$, but a *decrease* in Γ_{VH} for the depolarized scattering. The self-diffusivities of a rod, in general, decrease with the concentration because of the rod entanglement and hydrodynamic interaction, but the excluded volume effect in the polarized scattering largely cancels (and sometimes overcancels) the decrease. An example of the overcancellation is the PBLG data measured by Russo et al.

Using the present theory, we are now able to compute such quantities as the static structure factor [$g(\mathbf{k},0)$ and $g_{\mathrm{VH}}(\mathbf{k},0)$], initial decay rate [$\Gamma_{\mathbf{k}}$ and Γ_{VH}], profiles of the DLS spectrum [$g(\mathbf{k},t)/g(\mathbf{k},0)$ and $g_{\mathrm{VH}}(\mathbf{k},t)/g_{\mathrm{VH}}(\mathbf{k},0)$], and the magnitude of the excluded volume effect [$A(\nu)$ and $A_{\mathrm{VH}}(\nu)$] in the entire range of $\mathbf{k}$, t and ν for both polarized and depolarized scatterings.

As an example of the application of the theory, we reanalyzed [5] depolarized DLS data of rodlike polymers found in the literature [6]: Zero and Pecora measured the rotational diffusion coefficient $^*D_{\mathrm{r}}$ of PBLG in the semidilute regime. Assuming $^*D_{\mathrm{r}} = D_{\mathrm{r}}$ [or, in other words, $A_{\mathrm{VH}}(\nu) = 1$ for finite ν], they found the exponent x of the length dependence of $D_{\mathrm{r}} \propto L^{-x}$ to be $x = 7.6 \pm 0.6$. Though this value was 16% smaller than the theoretical x value of 9, they concluded the observed x was reasonably close to 9. We computed $A_{\mathrm{VH}}(\nu)$ in accordance with the experimental conditions used by Zero and Pecora, and divided the observed $^*D_{\mathrm{r}}$ with the computed $A_{\mathrm{VH}}(\nu)$ values to estimate the rotational self-diffusion coefficient D_{r}. The exponent x of D_{r} thus obtained is $x = 9.1 \pm 0.4$, which agrees with the theoretical x value within the experimental error.

It is concluded that the excluded volume effect on the DLS spectrum of rodlike polymers is by no means negligible, the magnitude of which we can estimate using the present theory. The dynamical structure factors $g(\mathbf{k},t)$ and $g_{\mathrm{VH}}(\mathbf{k},t)$ given in this paper will be useful in analyzing the DLS data of rodlike polymers in the dilute and semidilute regimes.

References:

[1] M. Doi, and S. F. Edwards, *The Theory of Polymer Dynamics* (Clarendon Press, Oxford, 1986).

[2] P. S. Russo, F. E. Karasz, and K. H. Langley, J. Chem. Phys. **80**, 5312 (1984).

[3] M. Doi, T. Shimada, and K. Okano, J. Chem. Phys. **88**, 4070 (1988).

[4] T. Maeda, Macromolecules **22**, 1881 (1989).

[5] T. Maeda, Macromolecules (in press).

[6] K. M. Zero, and R. Pecora, Macromolecules **15**, 87 (1982).

Pattern Formation in the Spatially Modulated Electrohydrodynamic Convection of Liquid Crystals

A. Ogawa[1], *K. Kawasaki*[1], *W. Zimmermann*[2], *and T. Kawakatsu*[1]

[1]Department of Physics, Faculty of Science, Kyushu University 33, Fukuoka 812, Japan

[2]Physikalisches Institut der Universität Bayreuth, D-8580 Bayreuth, Fed. Rep. of Germany

Abstract. We study two dimensional patterns and their stability in the spatially modulated electrohydrodynamic convection of liquid crystals through the analysis of an amplitude equation. We also propose a new equivalent spin model which shares some features of the amplitude equation in a certain case.

When an alternating voltage $V(\omega)$ applied across a thin layer of nematic liquid crystal is increased, instability occurs at a critical voltage $V_c(\omega)$. If the frequency ω is larger than a critical frequency ω_z, there appear convection rolls which are aligned normally to the undisturbed director vector $\vec{n}_0$. Below ω_z a direct instability to an oblique roll state appears. $(\omega_z, V_c(\omega_z))$ is called the Lifshitz Point[1].

We study the modulation of the roll patterns by applying a spatially periodic external field. The field has a characteristic wave number k in the same direction as $\vec{n}_0$ which we take as the x-axis. The wave number k is generally incommensurate to a natural wave number k_0 of a normal roll state and these two are related as $mk = n(k_0 + q)$ where m and n are integers and the misfit wave number q is assumed to be much smaller than k_0. Effects of the modulating field have been studied for a purely one-dimensional case in the region $\omega >> \omega_z$ by Coullet and Repaux[2]. They obtained phase diagrams including a locked state and quasi-periodic states, which they call a commensurate state and incommensurate states respectively.

Our interest is in the two dimensional patterns near the Lifshitz Point. They are described by the following amplitude equation[1],

$$\partial_t A = \{(\partial_x + iq)^2 - iz(\partial_x + iq)\partial_y^2 + W\partial_y{}^2 - \partial_y{}^4 + \mu - |A|^2\}A + \alpha A^{*n-1}, \quad (1)$$

where a parameter $|z| \leq 2$ and $W \propto \omega - \omega_z$. Parameter μ measures the deviation from the instability threshold, $\alpha (> 0)$ is the strength of the external field, and $A(x, y, t) = R\ \exp(i\theta)$ is the complex amplitude of a typical scalar quantity, e.g. the vertical component of velocity field, $v \propto A\ \exp\{i(k_0 + q)x\} + c.c.$. Using the phase dynamics method, we examined the stability of incommensurate states in the region where R can be approximated by a constant and obtained a stability criterion as

$$W < \{\frac{\pi^2}{8}\frac{2 - \kappa^2}{E(\kappa)^2} - 1\}zq, \quad (2)$$

Springer Proceedings in Physics, Vol. 52 Dynamics and Patterns in Complex Fluids
Editors: A. Onuki · K. Kawasaki

where $E(\kappa)$ is the complete elliptic integral of second kind, and κ satisfies the relation $16\alpha E(\kappa)^2\mu^{\frac{n-2}{2}} = n\pi^2 q^2\kappa^2$. The quantity in the curly bracket of (2) is positive. Thus when z and q are of the same sign, the modulation field acts in favor of the oblique roll structure against the normal roll structure.

In addition to the above analysis, we propose a new equivalent spin model which shares some features of the amplitude equation in a certain case. Our approach is related to that by Bak and von Boehm[3] in the ANNNI system. We assume the spin variable S_i at a site i to be $S_i = A(x_i, y_i, t)\exp\{i(k_0+q)x_i\}+c.c.$ for the slowly varying amplitude A. For example, we can show that the spin dynamical system with the following Hamiltonian (3) as the Lyapunov function reduces to (1) under suitable choice of J_{ij}'s, and h_2, h_3 and h_4 for $n = 3$.

$$H = \frac{1}{2}\sum_{i\neq j} J_{ij}S_iS_j + \sum_i (h_2{S_i}^2 + h_3{S_i}^3 + h_4{S_i}^4) \tag{3}$$

We need five exchange interactions as shown in the Fig.1. The model (3) exemplifies the variety of spin systems corresponding to the same amplitude equation (1).

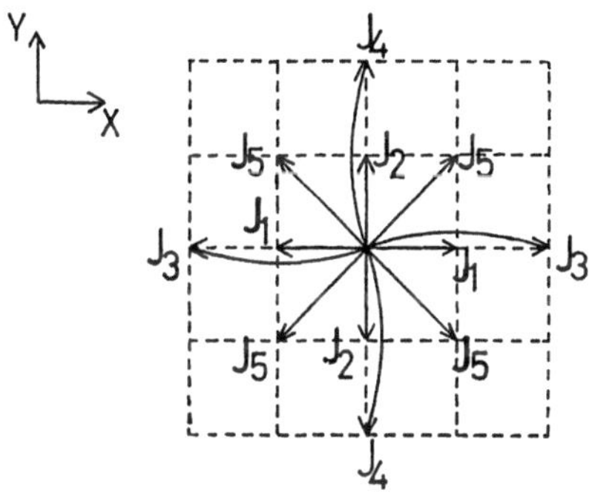

Fig.1 Five exchange interactions in a spin model (3)

References

1) W. Zimmermann, L. Kramer, Phys. Rev. Lett. **55**, 402 (1985)
W. Pesch and L. Kramer, Z. Phys. **B63**, 121 (1986).
E. Bodenschatz, W. Zimmermann and L. Kramer, J. Phys. France **49**, 1875 (1988).
2) P. Coullet and D. Repaux, Europhys. Lett. **3**, 573 (1987).
3) P. Bak and J. von Boehm, Phys. Rev. **B21**, 5297 (1980).

Theory of Shear Thinning in Liquid Crystal Polymers, for the Lower Shear Rate Regime

Y. Yamazaki [1,2,3], *A. Holz* [2], *S.F. Edwards* [3], *and H. Fujishiro* [1]

[1] Information Physics, Kyushu Institute of Technology, Iizuka 820, Japan
[2] Theoretical Physics, University of Saarland, D-6600 Saarbrücken, Fed. Rep. of Germany
[3] Cavendish Laboratory, Cambridge University, Madingley Road, Cambridge CB3 0HE, UK

Abstract. A nemato-hydrodynamic theory of shear thinning in solutions of lyotropic nematic liquid crystal polymers (LCP) consisting of semi-flexible polymer molecules is developed for the lower shear rate regime. Generalized power law behavior and crossover behavior are derived between the viscosity and the shear rate.

1. Introduction

The rheological behavior of a LCP system is usually rather complex [1,2]. For example, the typical shear rate($\dot{\gamma}$) dependence of the viscosity η of a LCP is found by Onogi and Asada[3]. The characteristic three regimes consist of two shear thinning regimes I, III and the Newtonian regime II. We study the lower shear rate regime I where domains of a LCP straighten out with increasing $\dot{\gamma}$.

2. Model and Fundamental Equations

We consider two media, an incompressible fluid with mass density ρ_0 and viscosity η_0 and a "porous" medium of domains consisting of polymer molecules. For the sake of simplicity we model the domains in a shear flow as a collection of rigid but permeable spheres with hydrodynamic permeability k being carried along by the flow.

The microscopic velocity field and pressure are determined by the Navier-Stokes equation and the incompressibility equation. The corresponding macroscopic field and pressure are derived by averaging those over a small space-time compared to a characteristic length and relaxation time. As the Reynolds number of a LCP is small the macroscopic fundamental equation is expressed by the Navier-Stokes equation without the nonlinear term and with the frictional force (exerted by the porous medium onto the fluid) within the frame of the mean field approximation.

3. Results and Conclusion

We consider a dilute solution of spheres containing n_p spheres per unit volume in the limit of non-interacting spheres. Let the solution be contained between two parallel walls situated at $y=\pm L$, and let the walls move with equal but opposite velocities($\pm GL$) along the x-axis. Suppose that the shear rate of the velocity field v_0 at a position r_i of a specific macromolecule is G_0. When the porous sphere is removed from the flow its presence yields a small correction to v_0 and G_0. Then the product of the viscosity η of the composite system and the shear rate G is equal to the external force $\eta_0\ G_0$ per unit area applied to the plates along the x-direction.

Springer Proceedings in Physics, Vol. 52 **Dynamics and Patterns in Complex Fluids**
Editors: A. Onuki · K. Kawasaki

In a theory of apparent viscosity of a LCP, the shear stress σ consists of the viscous stress σ' and the Ericksen stress σ_e due to the curvature elasticity of the LCP. The apparent viscosity is given by $\sigma/\dot{\gamma}$. Suppose now that $\sigma' \ll \sigma_e$ and that the change of distortion energy $\delta\varepsilon$ per unit volume is given by $\kappa_e \sigma_e^2/2$.
The first order calculation leads to

$$\eta = \eta_0/[(1-\delta)(1-B(\sigma)/B(\rho)+\delta] = (\kappa^{-2}-1)C/\dot{\gamma}$$

where

$$\delta(\rho) \equiv 1-bB(\rho),\ B(\rho) = [G_1(\rho)^{-1}+10\rho^{-2}]^{-1},\ b \equiv 5/2\Omega,\ \Omega \equiv (4\pi/3)R^3 n_p,$$

$$\sigma \equiv R/\sqrt{k},\ G_1(\rho) \equiv 1+3\rho^{-2}-3\rho^{-1}\mathrm{cth}\rho,\quad \kappa \equiv \sigma/\sigma_0,\ C \equiv \kappa/\sigma_0{}^2.$$

We consider a special limit $\delta \to 0$. The essentially dominant value of the viscosity of the LCP appears in the σ,ρ region near σ_0. There we can expand the denominator as

$$B(\sigma)-B(\rho) = (\sigma-\rho)B'(\rho)+\tfrac{1}{2}(\sigma-\rho)^2B''(\rho)+\cdots$$

where a prime denotes a derivative with respect to σ. Note that for $\rho=\sigma_0$ the first term vanishes and the second term dominates, whereas for $\rho \neq \sigma_0$ the first term is dominant and the second is negligible. If we take into account interactions among the spheres(e.g. from the first to the m-th nearest neighbor) or other characteristic parameters, the cases of the s-th derivative term of $B(\rho)$ dominant may be possible. There the vicosity of the system behaves as

$$\eta \sim \begin{cases} \dot{\gamma}^{-S/(S+1)} & S=1,2,3,\ldots \quad \text{for } \dot{\gamma}\to 0 \\ \text{const.} & \text{for } \dot{\gamma}\to\infty. \end{cases}$$

In general case of $\delta \neq 0$ we can derive the expressions similar to the above, which describe the crossover behavior and coincide with the behavior of isotropic polymer solution for $\delta=1$ in the absence of domains. The details are referred toin[4].

References

1. A.Ciferri,W.R.Krigbaum and R.B.Meyer,Eds.of "Polymer Liquid Crystals",(Academic Press,New York,London,1982)
2. K.F.Wissbrum,Lecture Notes in Mathematics,Vol.1063,pp.1, Ed. by J. L.Ericksen,(Springer,1984)
3. S.Onogi and T.Asada,Rheology,Vol.I,pp.127,Eds.G.Astarita et al., (Plenum Press,New York,1980)
4. Y.Yamazaki,A.Holz and S.F.Edwards,Phys.Rev.A

Hierarchical Discommensuration Pattern and Phase Transitions in the Electrohydrodynamic Convection of Liquid Crystals with a Periodic Substrate Potential

H. Mitani

Department of Physics, Fukuoka University of Education, Munakata, Hukuoka 811-41, Japan

1. Introduction

In recent years there has been considerable interest in the statistical mechanics of low-dimensional systems with competing interactions. A typical model of such systems is the Frenkel-Kontorova (FK) model or its extension, in which a one-dimensional array of particles interact with each other in the presence of a spatially periodic external potential [1] . Various concepts, such as discommensuration (DC), commensurate (C) lock-in, and fractal (or hierarchical) nature, have been introduced by the study of the models. In spite of these developments on the theoretical side, the experimental investigation of atomic systems with competitive interactions near the ground state has not been improved. The reasons are that: i) due to the microscopic size of atomic structure, we must see the system in the reciprocal space, ii) the temperature in the experiments is high enough to break the complex commensurate structures, iii) there may exist defects also in the substrate crystal within the commensurate periodicities. On the other hand, convective systems of liquid crystals with a periodic substrate voltage [2] are considered to be most adequate systems to observe the competitive phenomena, because the rolls which play the role of "atoms" have a semimicroscopic size, and thus the subtle structure is visible in real space, even if it does not have long-range order. Besides, its energy scale is larger than that corresponding to the possible temperature, so we can easily observe the structure at or near the ground state. In this paper, we interpret the experimental results of the liquid crystal system with a one-dimensional roll pattern, regarding it as a particle system with competitive interactions.

2. Hierarchical Discommensuration Structure

These competitive systems have the hierarchical discommensuration (DC) structure [3,4] . This structure is explained as follows: i) there is a periodic density wave of particles (which is DC), ii) there also exists a density wave of the DCs (which we name DC of DC), that is, the density of particles at the points which have equivalent phases on the "DC lattice" also have a periodic change; and so on. Such a hierarchical nature is mathematically represented by the continued fractional series of $r=b/a$, where b is the average roll or atom separation and a is the period of the substrate:

Springer Proceedings in Physics, Vol. 52 Dynamics and Patterns in Complex Fluids
Editors: A. Onuki · K. Kawasaki

$$r = b/a = k_0 + \frac{1}{k_1 +}\ \frac{1}{k_2 +}\ \frac{1}{k_3 + \dots} \ .$$

The physical meaning is as follows: the separation of the particles is approximately k_0 times the substrate period, the period of the DC lattice is approximately k_1 times the average particle separation, and the period of the DC of DC lattice is k_2 times the period of the DC lattice, and so on. (See Fig. 1.)

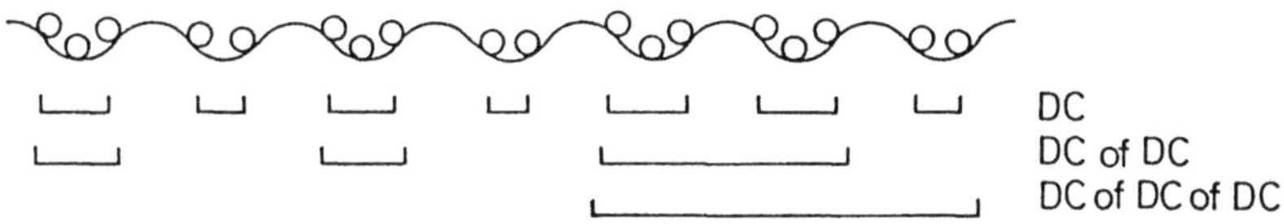

Fig.1 Hierarchical DC structure, where $r = 7/18$. The DC, DC of DC and DC of DC of DC appear according to the continued fraction. The dense regions in each period are indicated by brackets.

The experiment on the liquid crystal system by Low et al. was the first to show this hierarchical DC structure. Using the above continued fraction theory, we can interpret the observations of the complex roll density wave as a hierarchical DC structure. For example, the system with 5 roll pairs per 2 voltage periods is represented as $2/5 = 0 + 1/(2 + 1/2)$ (or $= 0 + 1/(3 - 1/2)$), which means the DC appears per 2 atoms, approximately, and the DC of DC appears per 2 DCs. The system with 7 roll pairs per 18 voltage periods is represented as $7/18 = 0 + 1/(3 + 1/(-2 + 1/(-3)))$, which means the DC appears per 3 rolls, approximately, and the DC of DC appears per 2 DCs, approximately, and the DC of DC of DC appears per 3 DC of DCs, approximately. This theory successfully explains the experimental results.

3. Phase Diagram

Hierarchical DC structure gives rise to the hierarchical energy structure versus structural parameters and "commensurate lock-in" [4] , because existence of DC means a deviation from a commensurate structure and it necessarily bears a DC energy (that is, defect energy). Further, it leads to hierarchical phase diagrams. Here, we explain the experimental phase diagram of the liquid crystal system by the one-dimensional particle model. We calculate the diagram by using the harmonic interaction potential (i.e. FK model) and the exponential one. We consider that the latter is more realistic for the present system, because there may be a strong tendency to maintain the radius of rolls and thus the potential profile may be asymmetric on the natural atomic distance. We show the phase diagram of the exponential potential (see Fig.2(a)). Lowe et al. also showed the change from a sharp phase transition between two phases to

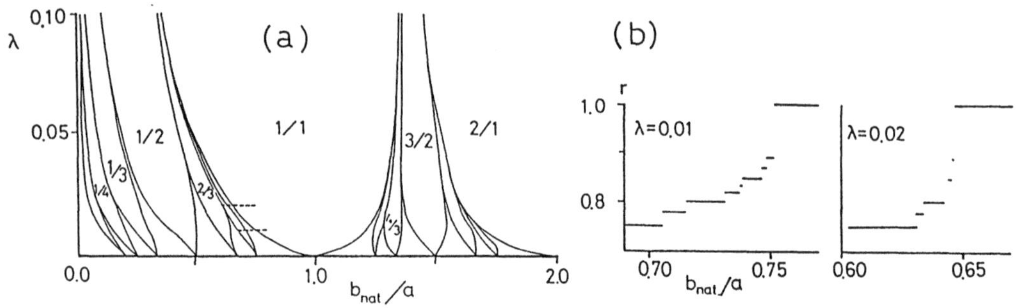

Fig. 2(a) The phase diagram on λ (substrate potential strength) and bnat.(natural particle distance), which is a function of the chemical potential, where the particle interaction potential is $\lambda \exp(-\xi r)$, and $\xi a=0.5$. Lowe showed the region $0.3<r<1.0$ experimentally. (b) Detailed graph of $r=b/a$ versus bnat./a near r=3/4 to 1/1. This region is indicated by broken lines in Fig.(a).

smooth (or devilish) changes near r=3/4 to 1/1. We interpret this phenomenon as there existing a multi critical point in the phase diagram accidentally (Fig.2(b)).

4. Discussion

Nelson et al. [5] , Coppersmith et al. [6] and Mitani [7] showed the finite temperature phase diagram of a monolayer system on a periodic substrate, which includes C phases, the incommensurate phase, and liquid phase. Using the discussion by Nelson, as far as the system of the rigid parallel roll pattern of the liquid crystal is concerned, there are only C phases with thermal defects, because the Kosterlitz-Thouless transition is absent and rotational or shear Lame constants are infinite for one-dimensional systems. However, such one-dimensional systems must be unstable when the temperature increases or some parameters change. Then the system will be two-dimensional with a one-dimensional nature, as the experimental results show. The microscopic calculation for this quasi-two-dimensional system is in progress.

References

1. S. Aubry, in Solitons and Condensed Matter Physics, ed. by A.R.Bishop, T.Schneider, Springer Ser. Solid-State Phys., Vol.8 (Springer, Berlin, Heidelberg 1979) p.264
2. M.Lowe et al. J.Fluid Mech. 173(1986)253
3. V.L. Pokrovski et al. J.Phys. C20 (1987) 1017
4. H.Mitani et al. J.Phys.C20 (1987) 1017
5. D.R.Nelson et al. Phys.Rev.B19(1979)2457
6. S.N.Coppersmith et al. Phys.Rev.B25(1982)349
7. H.Mitani, in Cooperative Dynamics in Complex Physical System, ed. by H.Takayama , Springer Ser. Syn., Vol.43 (Springer, Berlin, Heidelberg 1989) p54

Part IV

Polymers

Phase Transition and Self-Assembling Structures of Polymer Mixtures Under Shear Flow

T. Hashimoto, T. Takebe, and K. Fujioka

Department of Polymer Chemistry, Kyoto University,
Kyoto 606, Japan

Abstract We discuss self-assembling structures and phase transition of polymer systems under a Couette type shear-flow. Ternary systems with polymer A + polymer B + solvent, forming a semidilute solution of A/B mixtures, exhibit strongly elongated self-assembling structures under the shear and are brought back into single-phase state at a higher shear rate. The drop of critical temperature with shear is much larger than that for the critical mixture of simple liquids by three-to-four orders of magnitude. The ternary systems with off-critical compositions exhibit ellipsoidal droplets with an extremely narrow size distribution at a low shear rate and a shear-induced cluster-to-percolation transition at a higher shear-rate. Finally a trend completely opposite to the shear-induced homogenization will be presented for a binary mixture of polymer A + solvent.

1. Introduction

There are a number of interesting and fundamental observations for binary polymer mixtures of A and B without solvent (A/B) or with solvent (A/B/S) and solution of A (A/S) under shear flow. Homogeneous solutions (A/S) undergo phase separation under shear [1-8]. Such "shear-induced phase separation" has never been observed in small molecular systems. In contrast to the above phenomenon, phase separated A/B/S and A/B become homogeneous under shear [9-12]. This "shear-induced homogenization" and drop of critical temperature $\Delta T_c(S)$ as defined by,

$$\Delta T_c(S) \equiv T_c(0) - T_c(S) \quad , \tag{1}$$

$T_c(S)$ and $T_c(0)$ being critical temperatures at shear rate S and at S = 0, respectively, were also found for a binary mixture of simple liquids [13]. However, the size of the drop $\Delta T_c(S)$ is much bigger in the polymeric systems than in the small molecular systems as will be clarified in sec. 3.2. Whether or not the homogenized mixture under shear undergoes phase separation upon further increase of S may be an important question. It is important to understand the basic physics underlying these opposite phenomena systematically.

Another important observation at the strong segregation limit of the solutions and the mixtures is that shear promotes the burst of droplets but it also promotes coagulation of the burst droplets. The burst is driven by a viscous drag, and the coagulation is enhanced by a convection flow and driven by the interfacial tension. A balance of these two opposing forces results in steady-state domains as a self-assembling structure under the shear whose size R may be predicted by Taylor's theory [14] in the hydrodynamic regime,

$$R \sim \sigma / (\eta S) \quad , \tag{2}$$

where σ and η are the interfacial tension and viscosity of the liquids, respectively. Taylor's

Springer Proceedings in Physics, Vol. 52 Dynamics and Patterns in Complex Fluids
Editors: A. Onuki · K. Kawasaki

investigations were conducted on the single-drop system. Our system is more complex than Taylor's system, because it contains many droplets with a much smaller size studied by Taylor, as a result of large η in polymeric systems. Thus the validity of Taylor's theory to our systems needs to be confirmed (see sec. 4). Further questions arise on (i) how R varies with S when the domains are intricately percolated under shear and on (ii) what happens when S is further increased or the quench depth is further lowered so that the systems go effectively to a weak segregation limit (sec. 3). We will need a sound statistical physical approach for the nonequilibrium systems comprised of long chain molecules in order to answer these questions.

Our research generally involves studies of both rheology and phase transition (critical phenomenon) of mixtures (Fig. 1). Keys to describe rheology and phase transition will be (i) spectrum of concentration (composition) fluctuations and (ii) molecular orientation and deformation under shear. The facts that two kinds of relevant variables are generally involved and that the latter (ii) can be crucial, depending on S, are those unique to polymeric systems [7,8]. These two key quantities are really related to the structure factor of the mixture under shear flow which can be studied by the scattering method. Our goal is therefore to inter-relate these three basic physical phenomena, scattering, phase transition and rheology.

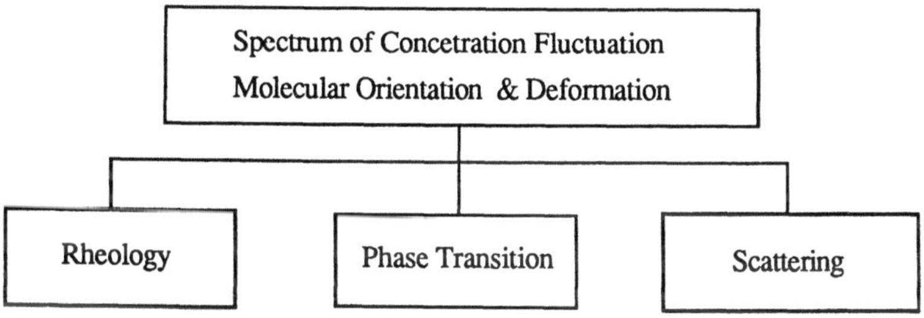

FIG. 1. Background in the study of mixtures under shear flow.

Intuitively we may think of two basic parameters underlying our system, as Onuki and Kawasaki employed in their theoretical work [15,16]: (i) S, i.e., the rate which we introduce externally to the system and (ii) growth rate or relaxation rate Γ_ξ for the concentration fluctuations and relaxation rate Γ_d for the molecular orientation and deformation, i.e., the rates intrinsic to the system under consideration. Note that only one kind of order-parameter, viz, the concentration fluctuation is important for the mixture of simple liquids. If $S<\Gamma_\xi$, the concentration fluctuations as well as orientation and deformation thermally relax before they are subjected to shear, and hence they are unaffected by shear, shear being insignificant. However, if $\Gamma_d \gg S > \Gamma_\xi$, the concentration fluctuations do not thermally relax before they are subjected to shear, and hence they are surely affected by shear, which affects the state of the mixtures and hence phase transition. However, the shear is still weak and insignificant for the molecular deformation. As S further increases and approaches Γ_d, the effect of molecular deformation is believed to play an important role [8].

2. Methods

We investigated ternary mixtures of polystyrene (PS), polybutadiene (PB) and dioctylphthalate (DOP), a common solvent for both PS and PB. The total amount of polymers in the systems are 3.0 and 3.3 wt%. They are semidilute solutions with concentration c satisfying $c/c^* \cong 2$,

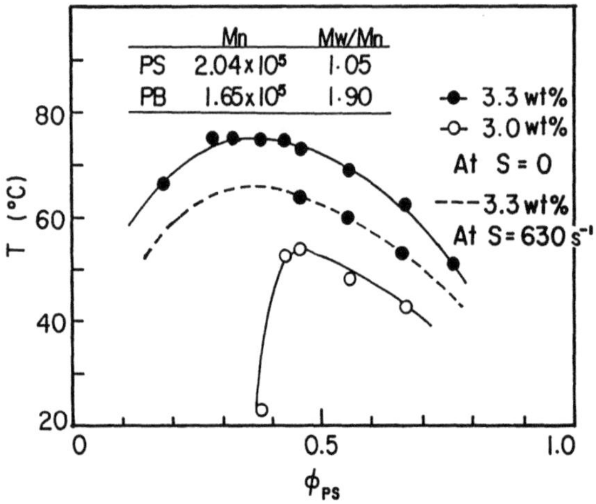

FIG. 2. Cloud-point curves for 3.0 and 3.3 wt % dioctylphthalate (DOP) solutions of PS/PB mixtures in quiescent state (solid lines) and at steady-state shear rate S = 630 s^{-1} (broken line). ϕ_{PS} is the volume fraction of PS in PS/PB mixture.

c^* being the overlap concentration [17]. The weight-average (M_w) and number-average molecular weight of the polymers (M_n) are shown in Fig. 2 which shows the cloud point curves measured at fixed c of 3.0 and 3.3 wt% in quiescent state (solid lines) or at S = 630 s^{-1} (broken line), dramatic drops of the cloud points were found as will be discussed in more detail in sec. 3.2. Rheological characteristics of the systems are as follows. Homopolymer solutions have about equal Newtonian viscosity, 1.0 and 3.0 poise for 3.0 wt% PS and PB solutions, respectively. 3.0 wt% polymer solution of the ternary mixtures with PS/PB = 50/50 wt% / wt% has the viscosity of ca. 2.0 poise at the shear-induced homogenized state.

We will also present some preliminary results on shear-enhanced concentration fluctuations for a semidilute solution of very high molecular weight polystyrene (UHPS) in DOP(sec. 5). The UHPS had $8.89 \times 10^6 \le M_n \le 1.36 \times 10^7$, and c = 1.0 wt%solution was used, $c/c^* \cong 10$.

We modified a cone-and-plate rheometer to study flow small angle light scattering (flow-SALS) under Couette type shear flow, the cone-and-plate being made of fused quartz [11,18]. The incident beam (He-Ne laser of 15 mW with wavelength of 632.8 nm in air) is irradiated through the velocity gradient and the SALS is detected in the directions parallel and perpendicular to the flow direction with a photodiode array, TV-camera, or photographic film, simultaneously with stress data [11,18].

3. Results for Critical Mixtures

3.1. Self-Assembling Structures under Shear Flow

Fig. 3 shows typical SALS patterns under steady state shear for a 3.0 wt% solution of 50/50 mixture of PS and PB at 10.8 K below cloud point $T_c(0)$ = 54.0 °C at S = 0 , the number below each pattern indicating S, and flow direction being in vertical [10]. The figure also includes the schematic models for the domains under flow. Fig. 4 shows corresponding

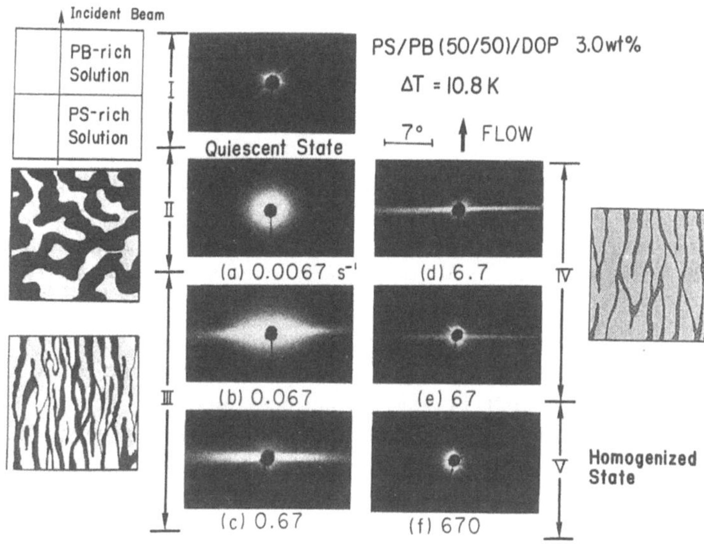

FIG. 3. Small-angle light scattering pattern for the 3.0 wt % PS/PB/DOP solution at 43.2 °C (quench depth $[T_C(0)-T] = 10.8$ K) under a quiescent state and under a steady-state Couette flow with S indicated beneath each pattern. The solution has a near critical composition (PS/PB 50/50 wt/wt). The angle mark indicates the scattering angle 7° in air. A schematic illustration of the domain structure is also included together with a classification of the shear effect into five regimes.

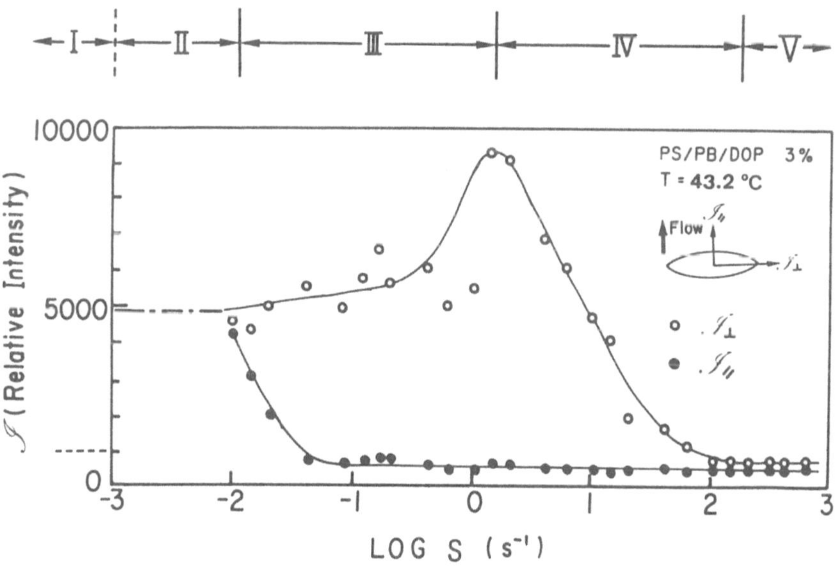

FIG. 4. Integrated intensities $\mathscr{I}_{//}$ and $\mathscr{I}_{\perp}$ parallel and perpendicular to the shear flow, respectively, as a function of shear rate S, and a classification of the state of the solution into five regimes.

integrated intensity parallel ($\mathcal{I}_{//}$) and perpendicular to the shear flow direction ($\mathcal{I}_{\perp}$) as a function of S [10]. The effect of flow seems to be classified into five regimes from the results shown in Figs. 3 and 4.

In regime I, the solution undergoes macroscopic phase separation, and the macroscopic interface is maintained. This regime occurs at S = 0 or at extremely low S. Since the incident beam propagates normal to the interface, the scattering is approximately equal to a sum of two homogeneous solutions, viz., PB-rich and PS-rich solution, and hence the scattered intensity is very weak as shown in Fig. 3 (the pattern for quiescent state, no appreciable intensity being detected) and in Fig. 4 (the intensity level of $\mathcal{I}_{//} = \mathcal{I}_{\perp} \cong 1000$ as drawn by the broken line).

In regime II strong SALS patterns appear at the scattering angles of a few degrees, indicating break-up of the macroscopic interface into the domain structure in an order of a few μm in size. Thus the balance of the two kinds of opposing physical forces, i.e., viscous force and thermodynamic force (interfacial tension) generates a unique self-assembling structure under the steady state shear flow. Since the pattern is approximately centersymmetric, the structure should have an approximately equal characteristic size parallel and perpendicular to the flow direction.

Moreover, all the domain structures in Regimes II to IV for the critical mixtures are expected to be intricately percolated, which will become clear by comparing the results with those for the off-critical mixtures (sec. 4). In a more rigorous sense percolation may exist definitely only in the direction of flow at a higher S.

In regime III, the patterns are elongated perpendicular to the flow direction as seen in the patterns (b) and (c) in Fig. 3, indicating that the domains are highly elongated parallel to the flow direction. The scattering distribution becomes anisotropic with $\mathcal{I}_{\perp} > \mathcal{I}_{//}$ as seen in Fig.4. It is quite impressive that the intensity $\mathcal{I}_{\perp}$ is suppressed at relatively low S down to the intensity level for the homogenized solution at a higher S in regime V. In regime IV at a still higher S, the elongated, streak-like patterns loose their intensity as seen in the patterns (d) and (e), which indicates that the concentration fluctuation $\Delta\phi(S)$ between the domains and matrix decreases with S due to an enhanced mixing of the unlike segments. This process is closely related to the drop of the cloud point $\Delta T_c(S)$ with S, as will be discussed in sec. 3.2.

In regime V at an even higher S, the SALS patterns disappear as seen in Fig. 3. The intensity levels of $\mathcal{I}_{\perp}$ and $\mathcal{I}_{//}$ in this regime are slightly lower than that of the homogeneous solution at S = 0 as seen in Fig. 4. This is a good implication of the system in regime V being in the homogenized state. It is worthy to note that the intensity parallel to shear flow is significantly suppressed in comparison with the intensity perpendicular to the flow direction even in this regime.

We found that the scattered intensity $I(q_y)$ perpendicular to shear flow in regimes III and IV obeys a scaling function as given by [19],

$$I(q_y) / I(0) = (1 + x^2)^{-2} \quad , \qquad (3)$$

where $x \equiv q_y \xi_{\perp}$ and $I(0) \sim \langle\eta^2\rangle \xi_{\perp}^2 \xi_{//}$. q_y is a component of scattering vector perpendicular to shear flow, $\xi_{\perp}$ and $\xi_{//}$ are S-dependent correlation lengths parallel and perpendicular to flow direction, respectively, $\langle\eta^2\rangle$ is the mean-squared refractive index fluctuations, and I(0) is the intensity at zero wavevector. The scaled structure factor corresponds to Debye's scattering formula [20] generalized for oriented two-phase systems [19]. Fig. 5 shows $\xi_{\perp}$ and $I(0)/\xi_{\perp}^2$ as a function of S in regimes III and IV, indicating that

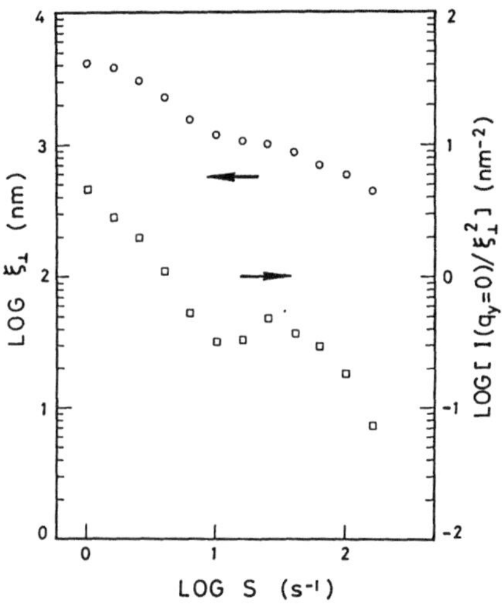

FIG. 5. Correlation length $\xi_\perp$ of the domains perpendicular to the shear flow and $I(q_y = 0)/\xi_\perp{}^2$ as a function of S in regimes III and IV.

$\xi_\perp$ decreases with S, approximately according to $S^{-1/3}$. The decrease of $I(0)/\xi_\perp{}^2$ may essentially reflect the decreases of $<\eta^2>$ and hence of $\Delta\phi$ with increasing S.

Fig. 6 shows the transient changes of the SALS patterns after cessation of the steady state shear flow in various regimes [21]. After cessation of the flow in regime III (part a) the streak-like patterns change into circular diffuse patterns with time, implying that the elongated domains under the flow become unstable in quiescent state to result in break-up into clusters. After cessation of the flow in regime V (part c) we observed the pattern change corresponding to spinodal decomposition (SD) after a temperature quench [22]. The shift of the "spinodal ring" toward smaller scattering wavevector q reflects the coarsening processes of the later stage SD. The cessation from the flow in regime IV (part b) gives a combination of the above two processes, i.e., the disappearance of the streak-like pattern, followed by evolution of the spinodal ring with some time lag. The disappearance of the streak should be closely related to Tomotika's instability [23]. However, Tomotika's stability analyses may have to be generalized to account for the complex nature of our system, especially for many-body interactions among viscous strands and time-evolution of the concentration fluctuations of the constituent polymers. Similar transient experiments [21] can be performed by shear-jumps from S in regime V to lower S's in various regimes where the spinodal ring is observed to appear and to be deformed to result in the steady-state domain structures corresponding to the patterns as shown in Fig. 3. The coarsening is found to be strongly anisotropic under the shear flow.

3.2. Shear-Induced Homogenization

The critical shear rate S_c for the shear-induced homogenization was determined as S above which $\mathcal{I}_\perp$ decreases to the level slightly lower than the intensity level of the homogeneous solution. The value S_c can be determined as a function of T or $[T_c(0)-T]$, and we found

$$S_c = (\text{const.}) \ [T_c(0)-T]^2 \quad , \tag{4}$$

where the constants are 1.4 and 4.9 for 3.0 and 3.3 wt % solutions, respectively [10,11,24]. The curve on S_c vs. T or $[T_c(0)-T]$ can be interpreted as a curve showing a drop of homogenization temperature T_c or ΔT_c as a function of S [11]. Fig. 7 shows the reduced

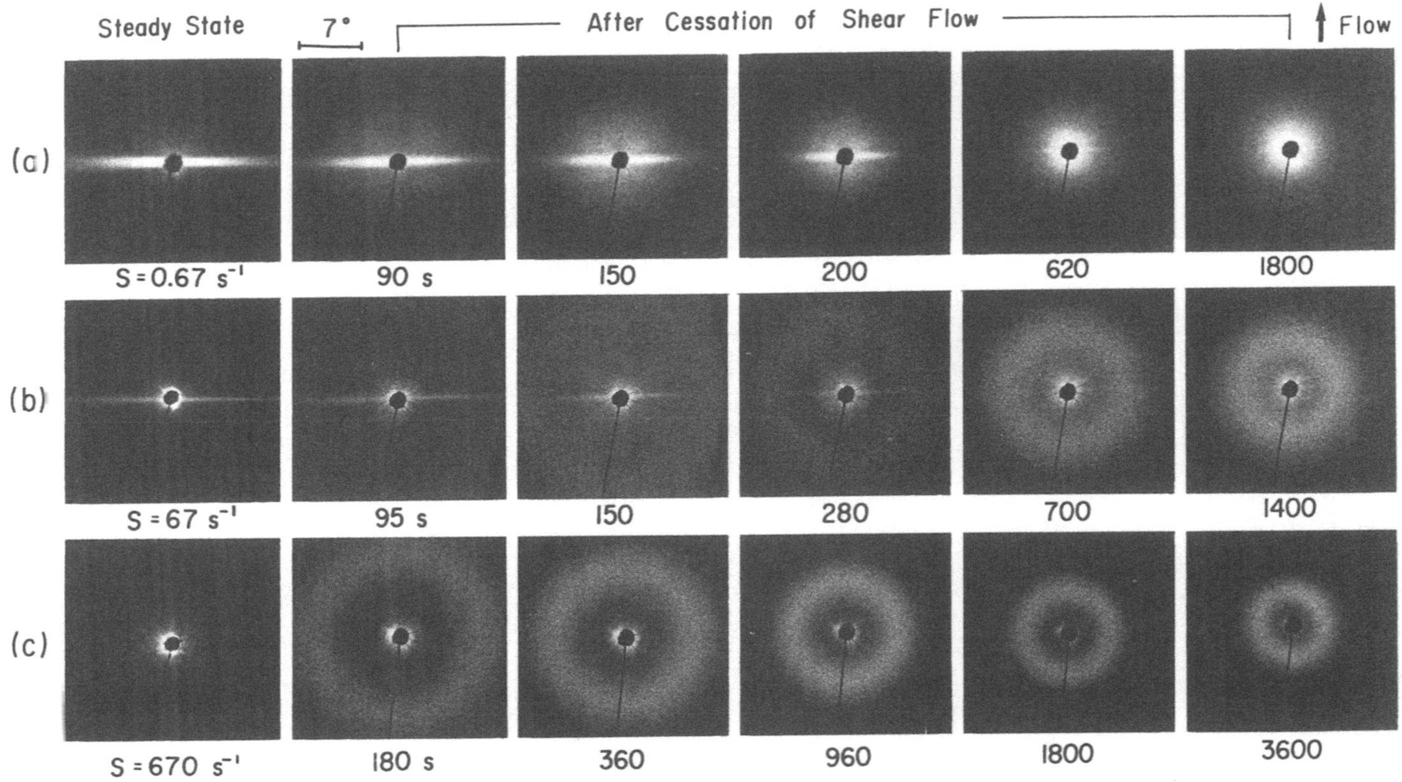

FIG. 6. Small-angle light scattering patterns under steady state shear (those in the left column) and time-evolution of the patterns after cessation of the steady-state shear at $S = 0.67\ s^{-1}$ in regime III (a), $67\ s^{-1}$ in regime IV (b) and $670\ s^{-1}$ in regime V (c). Time after the cessation (in seconds) is indicated beneath each pattern. The original flow direction is vertical.

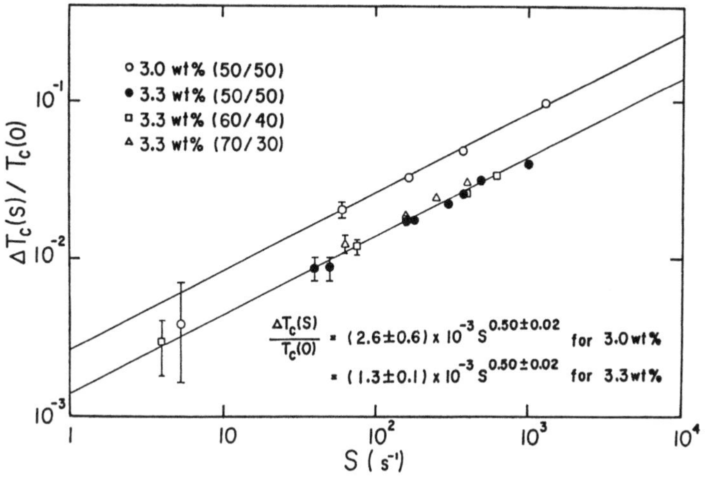

Fig. 7.Drop of the homogenization temperature $\Delta T_C(S)/T_C(0)$ as a function of shear rate S for the 3.0 and 3.3 wt % solutions of PS/PB/DOP. The 3.3 wt % DOP solutions of the PS/PB mixtures with the composition 50/50, 60/40 and 70/30 wt/wt gave identical results.

homogenization-temperature drop $\Delta T_C(S)/T_C(0)$ as a function of S in double logarithmic scale. The results lead to the following scaling laws;

$$\Delta T_C(S) / T_C(0) = P\, S^{0.50 \pm 0.02} \quad , \tag{5}$$

where the prefactors P's are $(2.6 \pm 0.6) \times 10^{-3}$ and $(1.3 \pm 0.1) \times 10^{-3}$, for 3.0 and 3.3 wt % solutions, respectively [11,19]. The difference in the prefactor primarily reflects the difference in solvent viscosity at $T_C(0)$. The semidilute solutions with c = 3.3 wt % having three different compositions gave approximately identical results.

We now compare our results on Eq. (5) with those reported by Beysens et al. on a critical binary mixture of simple liquids [13],

$$\Delta T_C(S) / T_C(0) = (5.9 \pm 0.7) \times 10^{-7}\, S^{0.53 \pm 0.03} \quad . \tag{6}$$

The scaling exponents are similar (0.50 vs. 0.53) but the prefactors are strikingly different, larger than that of the simple-liquid system by 3 to 4 orders of magnitude. The values $\Delta T_C(S)$'s for the polymer and simple-liquid systems are in the order of 10 K and 1 mK, respectively. We propose below a possible interpretation of this effect on the basis of the theory presented by Onuki et al.[15,16]. The theory developed for the critical binary mixture of simple liquids predicts well the experimental result of Eq. (6),

$$\Delta T_C(S)/T_C(0) = 0.0832\, \varepsilon\, \tau_\xi^{\ 1/3\nu}\, S^{1/3\nu} \quad , \tag{7}$$

where $\varepsilon = 4 - d$, d being spatial dimensionality, and ν is a critical exponent (0.63 for 3d kinetic Ising model and $1/3\nu = 0.53$). τ_ξ is the characteristic time for dynamics of the concentration fluctuations. When the dynamics is governed by hydrodynamic interactions, τ_ξ is given by

$$\tau_\xi = 16\,\eta\,\xi_0{}^3/k_B T \quad , \tag{8}$$

where ξ_0 is the microscopic length. The polymeric systems generally have a large τ_ξ compared with the small molecule systems, and hence it is not surprising at all that they have a large prefactor P and hence a large $\Delta T_c(S)$. We should bear in mind, however, that the dynamics of the concentration fluctuations in polymer systems may be different from that in simple liquid systems [10,11], as the hydrodynamic interaction tends to be screened out quite rapidly with increasing polymer concentration in the semidilute regime [17]. Moreover, polymer molecules change their conformations under shear flow, and then molecular deformation is another relevant variable in addition to the concentration fluctuations. This molecular deformation is mostly insignificant in small molecules but can be significant and hence a physical factor unique to high polymers. Therefore, application of Eq. (7) to our polymer systems is not straightforward. Nevertheless under a weak shear employed in this study where the molecular deformation is insignificant [10,11], the concentration fluctuation is only an important order parameter which controls the state of the mixtures, and hence $\Delta T_c(S)$ is primarily affected by dynamics of the concentration fluctuation. This is the basis of the qualitative interpretation that the large τ_ξ is responsible for the large $\Delta T_c(S)$, theoretical analysis of which is left as a future problem.

To understand the dynamics of the concentration fluctuations of our systems, we analyzed the mode-mode coupling contribution [11]. The net decay rate $\Gamma_{q,net}$ of the concentration fluctuations with wavenumber q is given [25,26] as a sum of the rate $\Gamma_{q,hydro}$ governed by the hydrodynamic interaction and the rate $\Gamma_{q,mol}$ governed by individual molecular motion, i.e., the rate in case when the hydrodynamic interaction is screened out,

$$\Gamma_{q,net} = \Gamma_{q,hydro} + \Gamma_{q,mol} \quad . \tag{9}$$

Eq. (9) gives a relationship among net diffusivity (D_{net}), and diffusivities controlled by hydrodynamic interaction (D_{hydro}) and individual molecular motion (D_{mol}),

$$D_{net} = D_{hydro} + D_{mol} \quad , \tag{10}$$

where $D_K \equiv \lim_{q\to 0} \Gamma_{q,K} / q^2$, and K is a subscript describing either "net", "hydro" or "mol". We can determine D_{net} from the analyses of early-stage SD after the cessation of shear from the regime V [11]. D_{hydro} is given by Kawasaki-Stokes formula [27],

$$D_{hydro} = k_B T/(6\pi\eta\xi) \quad , \tag{11}$$

where viscosity η can be measured, and correlation length ξ for the thermal concentration fluctuations can be determined also from the analyses of the early-stage SD, i.e., $\xi = 1/[\sqrt{2}q_m(0)]$, $q_m(0)$ being the wavenumber of the dominant mode of the growing fluctuation in the limit of t = 0 at the SD process. The mode-coupling contribution $MC \equiv D_{hydro}/D_{net}$ turns out to be between 0.2 and 0.5 for our systems. Hence the hydrodynamic interaction is suppressed, since $MC < 1$. Yet it is not completely screened out, still rather being effective, which gives the semiquantitative basis of applying Eq. (7) to our system [11].

The exponent 1/2 in Eq. (5) was proposed [11] to imply the mean-field behavior of our system, the dynamics of which is dominantly controlled by the individual molecular diffusion, i.e., $D_{mol} > D_{hydro}$. The mean-field behavior was confirmed also by analyses of the concentration fluctuations in regime V under the shear [11]. The scattered intensity perpendicular to the flow direction I (q_y) was found to obey Ornstein-Zernike formula,

$$I(q_y) = I(0)\,(1 + q_y{}^2\xi_\perp{}^2)^{-1} \quad , \tag{12}$$

where I(0) and $\xi_\perp$ were found to have S-dependence given by

$$I(0)^{-1} \sim S^{1/2} - S_c{}^{1/2} \quad , \tag{13}$$

and

$$\xi_\perp{}^{-2} \sim S^{1/2} - S_c{}^{1/2} \quad . \tag{14}$$

Eqs. (13) and (14) imply mean-field behavior, since $I(0)^{-1} \sim \Delta T_c(S) = T - T_c(S)$ and $\xi^{-2} \sim T - T_c(S)$, noting that $\Delta T_c(S) \sim S^{1/2}$.

4. Results for Off-Critical Mixtures

Fig. 8 shows the typical change in the scattering patterns with S for the off-critical mixture, 3.3 wt % DOP solution of a mixture of PS/PB 20/80 wt/wt at 62 °C, $[T_c(0)-T] = 9K$, where the flow direction is vertical, and S is indicated below each pattern. The pattern change with S

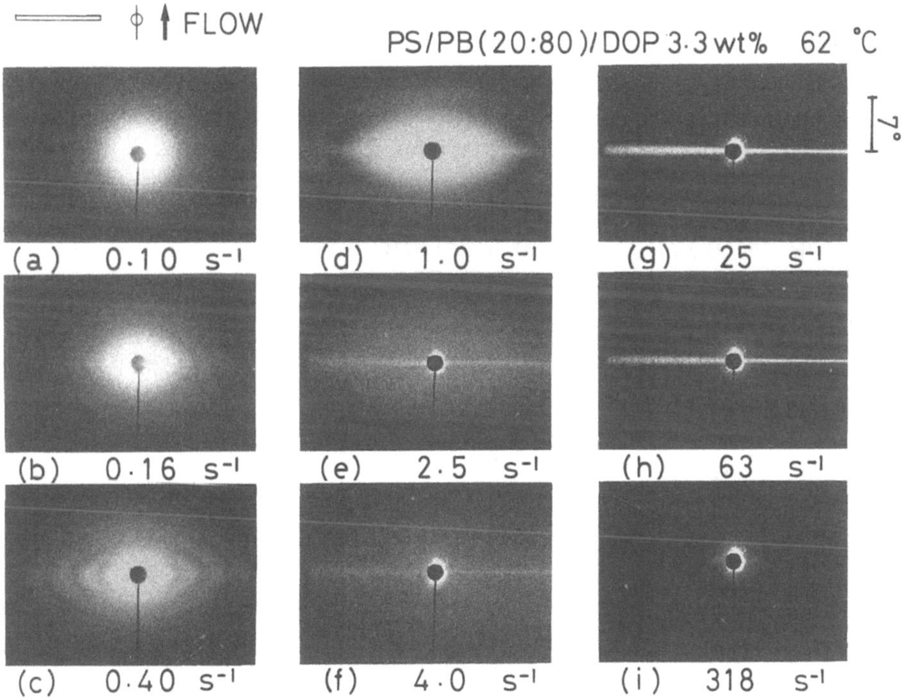

FIG. 8. Small-angle light scattering pattern for the 3.3 wt % PS/PB/DOP solution at 62 °C (a quench depth $[T_c(0)-T] = 9$ K) under a steady state Couette flow with S indicated beneath each pattern, and a classification of state of the solution into different regimes. The solution has an off-critical composition (PS/PB 20/80 wt/wt).

is similar to that for the critical mixture in the following points. (i) at a low S, the pattern is circular (regime II). It is transformed into the pattern elongated perpendicular to the flow direction and eventually into the sharp streak-like pattern (regime III). (ii) in regime IV, the streak-like pattern looses its intensity, and eventually the shear-induced homogenized state is attained in regime V. There are, however, striking differences between the off-critical and critical mixtures both in regime III and in crossover from regimes IV to V.

Off-critical mixtures exhibit remarkable hysteresis on $\mathcal{I}_\perp$ in the crossover behavior between IV to V ,i.e., $\mathcal{I}_\perp(S)$ obtained during increasing and decreasing cycles of S are different [24]. The difference of S_c observed in increasing and decreasing cycles of S (ΔS_c) was found to be insignificant for near critical mixtures but significant for off-critical mixtures, ΔS_c being larger (smaller) for smaller (larger) $[T_c(0)-T]$. The hysteresis effect may be a phenomenon relevant to the first-order phase transition, but has not been analyzed theoretically yet.

The off-critical mixtures exhibit a series of multiple-order scattering maxima at the lower shear-rate regime in regime III (defined as regime IIIa). Although the patterns (b) to (d) in Fig. 8 do not reproduce very well the multiple-order scattering maxima, the pattern at $S = 0.20\ s^{-1}$ revealed scattering maxima up to the 7th order at the positions exactly corresponding to the scattering maxima from the isolated particles of ellipsoid of revolution [28]. This evidence clearly implies the existence of oriented ellipsoidal particles as a flowing entity whose size distribution is quite narrow, as narrow as $\sigma_r/R_\perp \cong 0.04$ where σ_r is the standard deviation of the particle size perpendicular to the flow from the mean value $R_\perp$. Upon increasing S in regime IIIa, the scattering maxima shift toward higher scattering angles, and hence the scattering vectors of the first-order maximum, for example, parallel ($q_{m,//}$) and perpendicular to the flow direction ($q_{m,\perp}$) increase as shown in Fig. 9, where both $q_{m,\perp} \sim R_\perp{}^{-1}$ and $q_{m,//} \sim R_{//}{}^{-1}$ are proportional to S in accord with Taylor's theory [14], $R_{//}$ being the particle size parallel to the flow. In this regime the aspect ratio $\alpha \equiv R_{//}/R_\perp = q_{m,\perp}/q_{m,//} \cong 2$, independent of S.

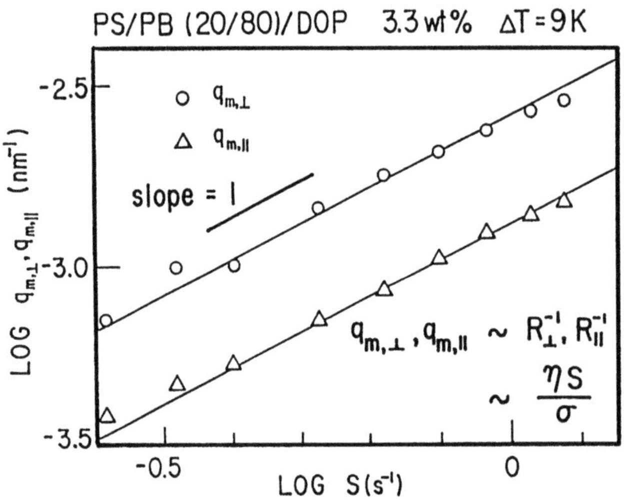

FIG. 9. Scattering vectors $q_{m,//}$ and $q_{m,\perp}$ parallel and perpendicular to the flow direction, respectively, for the first-order scattering maximum of the patterns in regime IIIa as a function of S for the off-critical mixtures shown in Fig. 8.

As S further increases in and beyond regime IIIa, the particles are divided into increasing number of decreasing volume with a fixed α, and hence the scattered intensity decreases. Simultaneously with this change, the horizontal streak starts to appear in the pattern and its intensity increases with increasing S, as seen in patterns (d) to (g) in Fig. 8. This implies that the numerous oriented small particles tend to form domains percolated parallel to the flow direction. As S increases, more particles are incorporated into the percolated domains and less particles exist as an isolated entity, as expected from decreasing particle scattering intensity and increasing streak-like scattering intensity. The higher shear-rate regime in III is defined here as IIIb where an interesting and striking change occurs in self-assembling structure, viz., the ellipsoidal droplets as a flowing entity undergo "shear-induced cluster-to-percolation transition".

5. Trend for Shear-Induced Phase Separation

We briefly discuss our preliminary study on shear-enhanced concentration fluctuation for 1.0 wt% DOP solution of UHPS, a semidilute solution with $c/c^* \cong 10$. The solution is homogeneous at 22 °C in quiescent state. As shown in Fig. 10, the scattered intensity $\mathscr{I}_\perp$ is very weak at a low S and equal to that at S = 0. Hence at a low S the solution is homogeneous. However, upon further increasing S above $S_c \cong 64\ s^{-1}$, $\mathscr{I}_\perp$ suddenly increases dramatically. This clearly implies shear-enhanced concentration fluctuations and possibly shear-induced phase-separation.

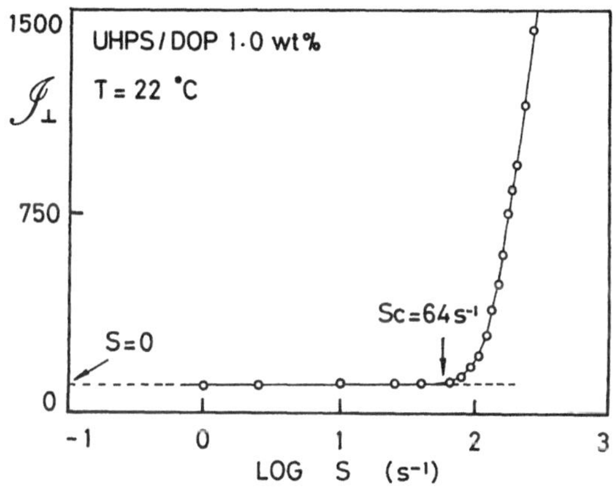

FIG. 10. Integrated intensity $\mathscr{I}_\perp$ as a function of the shear rate S for the 1.0 wt % UHPS/DOP solution at 22 °C.

Fig. 11 compares self-assembling structures in the two-phase state for the two kinds of systems, the ternary system PS/PB/DOP (critical mixture) showing shear-induced homogenization (patterns a and b) and the binary system UHPS/DOP showing shear-enhanced concentration fluctuations (patterns c and d). In the single-phase state both systems show no appreciable scattering (patterns b and c). When S is lowered, the critical mixture PS/PB/DOP shows a highly elongated domain structure as a steady-state self-assembling structure as discussed in sec. 3.1 (pattern a). The scattering pattern (d) obtained for the binary mixture upon increasing S is strikingly different from that shown in pattern (a), i.e., pattern (d) exhibited an approximately circular diffuse pattern, while the pattern (a) exhibited a sharp streak-like pattern. This difference implies that the self-assembling steady-state structure for

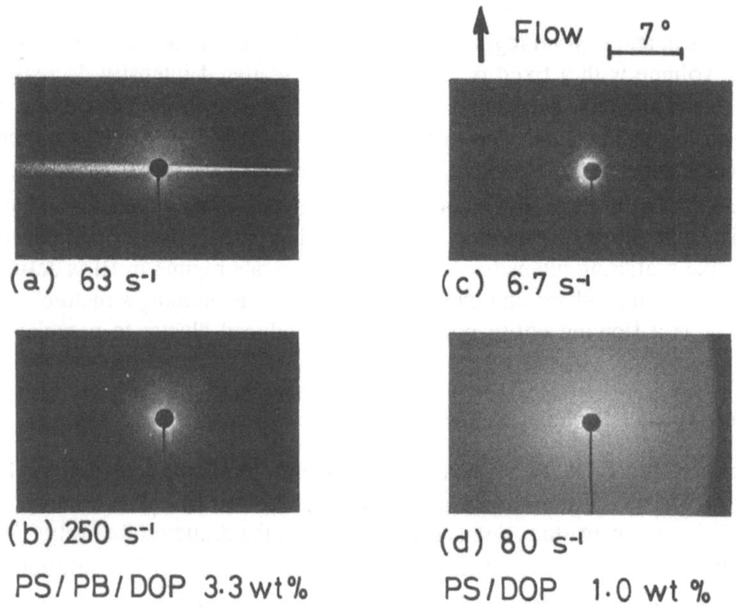

FIG.11. Comparison of the steady-state small-angle light scattering patterns for the PS/PB/DOP solution showing shear-induced homogenization (the patterns (a) and (b) at S = 63 and 250 s^{-1}, respectively) and those for the UHPS/DOP solution showing shear-enhanced concentration fluctuations (the patterns (c) and (d) at S = 6.7 and 80 s^{-1} respectively). Note the striking difference between the patterns (a) and (d).

the binary mixture is quite different from that for the ternary mixture. No appreciable depolarized component of scattered light was observed for either system.

Acknowledgments

The authors thank Professors A.Onuki and T.Ohta for stimulating discussion.This work was supported in part by a Grant-in-Aid for Scientific Research in Priority Areas "New Functionality Materials Design, Preparation and Control" (No.63604558) from the Ministry of Education, Science and Culture, Japan, and by the Mitsubishi Foundation.

References

1) G. Ver Strate and W. Philippoff, J. Polym. Sci. Polym. Lett. Ed., **12**, 267 (1974).
2) C.Rangel-Nafaile, A.B. Metzner, and K.F. Wissbrum, Macromolecules **17**, 1187 (1984)
3) J.R. Schmidt and B.A. Wolf, Colloid. Polym. Sci. **257**, 1188 (1979).
4) H. Krämer and B.A. Wolf, Makrmol. Chem. Rapid Commun. **6**, 21 (1985); H. Krämer-Lucas, H. Schenck, and B.A. Wolf, Makromol. Chem. **189**, 1613 (1988); ibid. **189**, 1627 (1988).
5) B.A. Wolf, Macromolecules **17**, 615 (1984).

6) G. Marrucci, Trans. Soc. Rheol. **16,** 1321 (1972).
7) E. Helfand and G.H. Fredrickson, Phys. Rev. Lett. **62**, 2468 (1989).
8) A. Onuki, Phys. Rev. Lett. **62**, 2472 (1989).
9) A. Silverberg and W. Kuhn, J. Polym. Sci. **13**, 21 (1954).
10) T. Hashimoto, T. Takebe, and S. Suehiro, J. Chem. Phys. **88**, 5874 (1988); T. Takebe and T. Hashimoto, Polym. Commun. **29**, 227 (1988).
11) T. Takebe, R. Sawaoka, and T. Hashimoto, J. Chem. Phys. **91**, 4369 (1989).
12) F.B. Cheikh Larbi, M.F. Malone, H.H. Winter, J.L. Halary, M.H. Leviet, and L. Monnerie, Macromolecules **21**, 3534 (1988).
13) D. Beysens, M. Gbadamassi, and L. Boyer, Phys. Rev. Lett. **43**, 1253 (1979); D. Beysens and M. Gbadamassi, J. Phys. (Paris) Lett. **40,** L565 (1979).
14) G.I. Taylor, Proc. R. Soc. London, Ser. A **138**, 41 (1932); **146,** 501 (1934).
15) A. Onuki and K. Kawasaki, Ann. Phys. (N.Y.) **121,** 456 (1979); Suppl. Progr. Theor. Phys. **64**, 436 (1978); Phys. Lett. **72A**, 233 (1979).
16) A. Onuki, K. Yamazaki, and K. Kawasaki, Ann. Phys. (N.Y.) **131**, 217 (1981).
17) P.G. de Gennes, *Scaling Concepts in Polymer Physics*, Cornell Univ. Press, Ithaka, N.Y., 1979.
18) T. Hashimoto, T. Takebe, and S. Suehiro, Polym. J. **18**, 123 (1986).
19) T. Takebe. K. Fujioka, R. Sawaoka, and T. Hashimoto, unpublished results.
20) P. Debye and A.M. Bueche, J. Appl. Phys. **20**, 518 (1949): P. Debye, H.R. Anderson, and H. Brumberger, J. Appl. Phys., **28**, 679 (1957).
21) T. Takebe and T. Hashimoto, Polym. Commun. **29**, 261 (1988).
22) T. Hashimoto, Phase Transitions **12**, 47 (1988).
23) S. Tomotika, Proc. R. Soc. London Ser. A, **150**, 322 (1935);**153**, 302 (1936).
24) T. Takebe, K. Fujioka, and T. Hashimoto, unpublished results.
25) K. Kawasaki, in *Phase Transitions and Critical Phenomena*, ed. by C. Domb and M.S.Green, Academic, N.Y. 1976, Vol. 5a.
26) R. Ferrell, Phys. Rev. Lett. **24**, 1169 (1970).
27) K.Kawasaki, Ann.Phys. (N.Y.) **61**,1(1970)
28) T. Hashimoto and K. Fujioka, unpublished results.

Effects of Viscoelasticity on Polymer Diffusion

M. Doi

Department of Applied Physics, Faculty of Engineering, Nagoya University, Chikusa, Nagoya, Japan

Abstract. It is demonstrated that the coupling between diffusion and macroscopic flow becomes important in various situations of polymer solutions. A phenomenological constitutive equation which describes this effect is proposed. As an application, two problems are discussed: (i) viscoelastic effect in the swelling process of polymer solutions, and (ii) concentration gradient in the rheometrical shear flow in parallel plate and cone-and-plate geometries.

1. Introduction

The problem I shall discuss here is the coupling between diffusion and macroscopic flow in polymer solutions. This problem is related to many problems which have been discussed recently.[1-11]

(i) Concentration gradient in inhomogeneous flow: Consider a polymer solution sheared between parallel disks (see Fig.1). Will the concentration of the polymer remain uniform? It has been argued[1-3] that the polymer concentration in the middle should be higher than that near the edge since the polymers near the edge are stretched by the shear flow and thus tend to move toward the center to reduce the elastic energy. Is this argument correct? Curiously enough, no experiment has been done to check this argument.

(ii) Polymer dissolution: Consider that a polymer solvent is added on top of a polymer solution in a tube (see Fig.2). The polymer

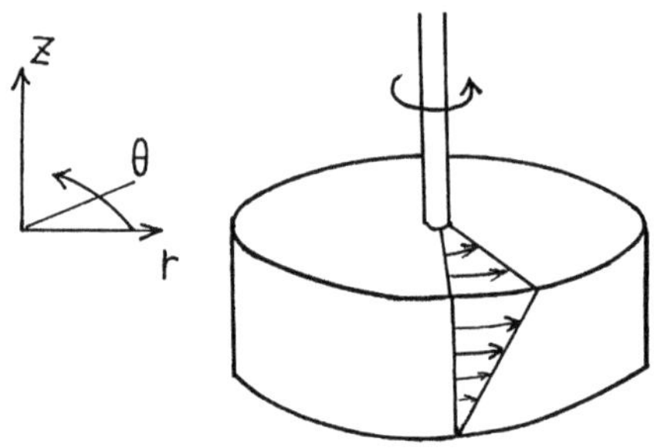

Fig.1 Shear flow between parallel disks. The velocity profile is given by $v_r=0$, $v_\theta=crz$ and $v_z=0$

Springer Proceedings in Physics, Vol. 52 Dynamics and Patterns in Complex Fluids
Editors: A. Onuki · K. Kawasaki

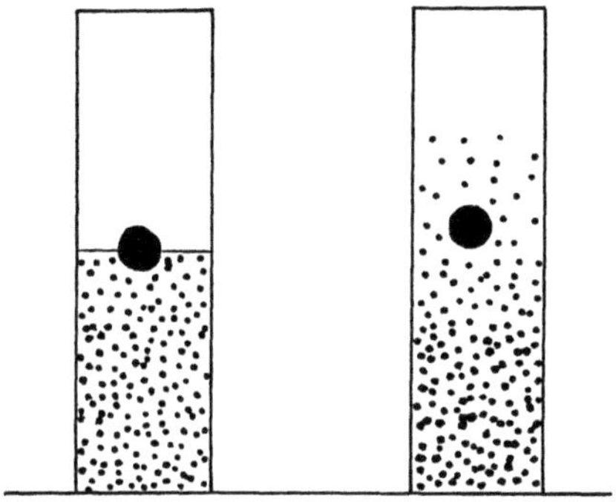

Fig.2 Polymer dissolution. Initially a solvent is put on top of a polymer solution in a tube, then the polymer will swell toward the solvent region.

will move toward the dilute region by diffusion. Does this process induce a macroscopic flow? For example, if one puts a small particle in it, will it move? For a mixture of simple liquids, the particle will not move, but if one component is a high molecular weight polymer (such as a gel), the particle will move. What is the physical origin of such a distinction? This problem is also related to the concentration fluctuation in polymer solutions [4] and also to non-Fickian diffusion in polymer systems [5-7].

(iii) Biphasic shear flow: Polymer solutions in poor solvent become phase separated when temperature is lowered. If we apply a shear flow, will the phase boundary change? If it does, what is the condition for the two phases to coexist? This problem has been discussed over years (see [8,9] for a review), and there has been a resurgence of interest recently [10,11].

These problems are not easy to answer. Indeed some have been discussed over many years. Here I will propose a phenomenological constitutive equation which I hope will be able to handle these questions. The equation describes the coupling between the macroscopic flow and polymer diffusion. Such an equation has been lacking in the conventional constitutive equation in polymer rheology, which makes the assumption that the system is uniform.

The constitutive equation is obtained by a generalization of two well-established phenomenological theories, i.e., the polymer rheology [12] and the critical fluid dynamics [13]. It is not entirely original: a prototype of the idea can be traced in many early theories. I am particularly inspired by [5,16] and [17]. Since this approach is basically phenomenological, there is an uncertainty in its validity, which must be checked by experiments. Despite the uncertainty, I believe it is useful to give a testable prediction at this stage.

2. Two Fluid Model

In order to describe the coupling between diffusion and macroscopic flow, I use the two fluid model. Let $\mathbf{v}_p(\mathbf{r},t)$ and $\mathbf{v}_s(\mathbf{r},t)$ be the macroscopic velocity of polymer and solvent at point $\mathbf{r}$ and time t. Let $\phi(\mathbf{r},t)$ be the volume fraction of the polymer. The volume averaged velocity, or the macroscopic velocity $\mathbf{v}(\mathbf{r},t)$ is given by

$$\mathbf{v}(\mathbf{r},t)=\phi\mathbf{v}_p+(1-\phi)\mathbf{v}_s. \tag{2.1}$$

I assume that the solution is incompressible

$$\nabla\cdot\mathbf{v}=0 \tag{2.2}$$

and that the molar volume of each component is constant. Thus the equation of the conservation of polymer mass is written as

$$\frac{\partial\phi}{\partial t} = -\nabla\cdot(\phi\mathbf{v}_p). \tag{2.3}$$

To derive the equations that determine $\mathbf{v}(\mathbf{r},t)$ and $\mathbf{v}_p(\mathbf{r},t)$, I shall use the variational principle of Rayleigh [14]. This variational principle has been extensively used in the dynamics of polymer solutions and suspensions [15]

Consider a set of macroscopic variables $x_1,x_2,\ldots=\{x_i\}$, and denote their time derivatives as $\dot{x}_1,\dot{x}_2,\ldots=\{\dot{x}_i\}$. The variational principle states that $\{\dot{x}_i\}$ is determined by the condition that the function

$$K=\frac{1}{2}W + \dot{A} \tag{2.4}$$

be minimized at each instant with respect to $\{\dot{x}_i\}$. Here W represents the energy dissipation in the system and is given by a quadratic function of $\{\dot{x}_i\}$,

$$W=\sum_{i,j}\zeta_{ij}\dot{x}_i\dot{x}_j \tag{2.5}$$

and $\dot{A}$ represents the time variation of the free energy A: $\dot{A}$ is written by the thermodynamic forces F_i conjugate to x_i;

$$\dot{A} = -\sum_i F_i\dot{x}_i\ . \tag{2.6}$$

The condition that $(1/2)W+\dot{A}$ be minimized with respect to $\dot{x}_i$ gives

$$\sum_j \zeta_{ij}\dot{x}_j = F_i. \tag{2.7}$$

This gives the time evolution equation of $\{x_i\}$.

Now in the present problem, $\{\dot{x}_i\}$ corresponds to $\mathbf{v}_p(\mathbf{r},t)$ and $\mathbf{v}(\mathbf{r},t)$. For the two fluid model, I assume that the energy dissipation function is given by

$$W = \int d\mathbf{r} \left\{ \zeta(\phi)\left[\mathbf{v}_p(\mathbf{r},t)-\mathbf{v}_s(\mathbf{r},t)\right]^2 + \eta_s(\nabla\mathbf{v}:\nabla\mathbf{v}) \right\}. \tag{2.8}$$

The first term is the energy dissipation caused by the relative motion of polymer and solvent (ζ being called the friction constant per unit volume). The second term represents the energy dissipation caused by the velocity gradient of the macroscopic flow. In the critical fluid dynamics, η_s represents the viscosity of the normal part, but for polymer solutions, η_s should be the viscosity of the solvent. (Strictly speaking, this term includes a subtle problem. One might consider that the second term should be $\eta_s(\nabla\mathbf{v}_s:\nabla\mathbf{v}_s)$ or $\eta_s(1-\phi)(\nabla\mathbf{v}_s:\nabla\mathbf{v}_s)$ since the energy dissipation in polymer solution is mainly caused in the solvent. However, I will not dwell on this problem here since the term is usually small in entangled polymer systems and will be eventually neglected in the present theory.)

Let $\mathbf{F}_p$ and $\mathbf{F}_s$ be the forces acting on polymer and solvent in unit volume respectively, then $\dot{A}$ is written as

$$\dot{A} = -\int d\mathbf{r}\, (\mathbf{v}_p\cdot\mathbf{F}_p + \mathbf{v}_s\cdot\mathbf{F}_s). \tag{2.9}$$

If the explicit form of the free energy is known, $\mathbf{F}_p$ and $\mathbf{F}_s$ are obtained by this equation, as will be demonstrated later.

In minimizing $K=(1/2)W+\dot{A}$ with respect to $\mathbf{v}_p$ and $\mathbf{v}$, we have to remember that $\mathbf{v}$ satisfies the incompressible condition (2.2). This constraint can be accounted for by Lagrange's method of unknown constants. The function to be minimized is

$$\begin{aligned}\tilde{K} &= \frac{1}{2}W + \dot{A} - \int d\mathbf{r}\, p(\mathbf{r},t)\nabla\cdot\mathbf{v} \\ &= \int d\mathbf{r} \left\{ \frac{\zeta}{2(1-\phi)^2}(\mathbf{v}_p-\mathbf{v})^2 + \frac{\eta_s}{2}(\nabla\mathbf{v}:\nabla\mathbf{v}) \right. \\ &\qquad \left. - \left[\mathbf{F}_p - \frac{\phi\mathbf{F}_s}{1-\phi}\right]\cdot\mathbf{v}_p - \frac{\mathbf{F}_s}{1-\phi}\cdot\mathbf{v} - p\nabla\cdot\mathbf{v} \right\},\end{aligned} \tag{2.10}$$

where $\mathbf{v}_p$ and $\mathbf{v}$ are chosen as independent variables. The condition $\delta\tilde{K}/\delta v_p = \delta\tilde{K}/\delta v = 0$ gives

$$\frac{\zeta}{(1-\phi)^2}(\mathbf{v}_p-\mathbf{v}) = \mathbf{F}_p - \frac{\phi\mathbf{F}_s}{1-\phi} \tag{2.11}$$

and

$$-\frac{\zeta}{(1-\phi)^2}(\mathbf{v}_p-\mathbf{v}) - \eta_s\nabla^2\mathbf{v} = \frac{\mathbf{F}_s}{1-\phi} - \nabla p\ . \tag{2.12}$$

The unknown function $p(\mathbf{r},t)$ is determined by the incompressible condition(2.2). From eqs.(2.11) and (2.12), it follows

$$\eta_s \nabla^2 \mathbf{v} - \nabla p + (\mathbf{F}_p + \mathbf{F}_s) = 0 \ . \tag{2.13}$$

This equation represents the condition of mechanical balance. (Notice that the inertia term $\rho D\mathbf{v}/Dt$ has been neglected in the present theory.) On the other hand, eq.(2.11) describes the relative motion between polymer and solvent.

So far the formulation is quite general for the two fluid model. I will now consider explicit forms of $\mathbf{F}_p$ and $\mathbf{F}_s$ for some examples.

3. Thermodynamic Forces

3.1 Critical fluid dynamics

First I consider the critical fluid dynamics, where both components are of low molecular weight. (Thus p and s in this section simply represent each component of the mixture.) The free energy of such a system is a functional of ϕ, and may be written as [13]

$$A = \int d\mathbf{r} \left[f(\phi(\mathbf{r})) + \frac{1}{2} K (\nabla\phi)^2 \right] . \tag{3.1}$$

Hence $\dot{A}$ is calculated as

$$\dot{A} = \int d\mathbf{r}\, \frac{\delta A}{\delta \phi} \cdot \dot{\phi} = \int d\mathbf{r}\, \frac{\delta A}{\delta \phi} [-\nabla \cdot (\phi \mathbf{v}_p)] = \int d\mathbf{r}\, \phi \mathbf{v}_p \cdot \nabla \left(\frac{\delta A}{\delta \phi} \right) . \tag{3.2}$$

Thus the thermodynamic force is given by

$$\mathbf{F}_p = - \phi \nabla \left(\frac{\delta A}{\delta \phi} \right) = - \phi \nabla \left(\frac{\partial f}{\partial \phi} - K \nabla^2 \phi \right) , \qquad \mathbf{F}_s = 0 \ . \tag{3.3}$$

Eqs.(2.11) and (2.13) are then written as

$$\mathbf{v}_p - \mathbf{v} = - \frac{(1-\phi)^2 \phi}{\zeta} \nabla \left(\frac{\delta A}{\delta \phi} \right) \tag{3.4}$$

$$\eta_s \nabla^2 \mathbf{v} - \nabla p - \phi \nabla \left(\frac{\delta A}{\delta \phi} \right) = 0 \ . \tag{3.5}$$

Eqs.(3.4) and (2.3) give

$$\frac{\partial \phi}{\partial t} + \mathrm{v} \nabla \phi = \nabla \lambda \phi \cdot \nabla \left(\frac{\delta A}{\delta \phi} \right) \tag{3.6}$$

where

$$\lambda = \frac{(1-\phi)^2 \phi}{\zeta} \ . \tag{3.7}$$

Eqs.(3.4) and (3.5) are the basic equations in the critical fluid dynamics [13].

The force $\mathbf{F}_p$ can be written as a divergence of a tensor [17]:

$$\mathbf{F}_p = \nabla \cdot \boldsymbol{\sigma}_p \tag{3.8}$$

with

$$\boldsymbol{\sigma}_p = - \Pi \mathbf{I} + \boldsymbol{\sigma}_K \tag{3.9}$$

where

$$\Pi = \phi \frac{\partial f}{\partial \phi} - f \tag{3.10}$$

denotes the osmotic pressure of the system and

$$\sigma_{K\alpha\beta} = -\left[\frac{1}{2}K(\nabla\phi)^2 + K\phi\nabla^2\phi\right]\delta_{\alpha\beta} - K\frac{\partial\phi}{\partial r_\alpha}\frac{\partial\phi}{\partial r_\beta} \tag{3.11}$$

denotes the stress arising from the interfacial tension.

The stress σ_K contributes to the macroscopic stress. Indeed, the mechanical balance equation (3.5) is written as

$$\nabla\cdot\left\{\eta_s\left[\nabla \mathbf{v} + (\nabla \mathbf{v})^{+}\right] - p\mathbf{I} - \Pi\mathbf{I} + \boldsymbol{\sigma}_K\right\} = 0 . \tag{3.12}$$

This indicates that the tensor

$$\boldsymbol{\sigma} = \eta_s\left[\nabla \mathbf{v} + (\nabla \mathbf{v})^{+}\right] - p\mathbf{I} - \Pi\mathbf{I} + \boldsymbol{\sigma}_K \tag{3.13}$$

denotes the macroscopic mechanical stress. Notice that $\boldsymbol{\sigma}_K$ affects the stress $\boldsymbol{\sigma}$, while the osmotic pressure Π has no effect on the macroscopic stress since Π in eq.(3.13) can be completely absorbed in the function $p(\mathbf{r},t)$.

3.2 Gels

In the solution of simple liquids, the free energy can be expressed as a functional of the volume fraction $\phi(\mathbf{r})$ only. In polymeric systems, however, this is not true since the conformational entropy of the polymer chain has a large contribution to the free energy. This becomes particularly clear for swollen gels, so I shall first discuss these.

A point on a gel can be denoted by the position vector $\mathbf{r}_0$ at a certain reference state. Let $\mathbf{r}(\mathbf{r}_0,t)$ be the position vector of this point at time t. The function $\mathbf{r}=\mathbf{r}(\mathbf{r}_0,t)$ may be inverted to give $\mathbf{r}_0 = \mathbf{r}_0(\mathbf{r},t)$. The deformation of the gel is described by the tensor:

$$E_{\alpha\beta}(\mathbf{r},t) = \left.\frac{\partial r_\alpha}{\partial r_{0\beta}}\right|_{\mathbf{r}_0=\mathbf{r}_0(\mathbf{r},t)} . \tag{3.14}$$

(Notice that $\mathbf{E}$ is expressed as a function of $\mathbf{r}$ and t.) The polymer volume fraction is expressed by $\mathbf{E}(\mathbf{r},t)$ as

$$\phi(\mathbf{r},t)=\phi_0/\det(E_{\alpha\beta}(\mathbf{r},t)) \tag{3.15}$$

where ϕ_0 is the volume fraction at the reference state. Also one can show

$$\dot{E}_{\alpha\beta} \equiv \frac{\partial}{\partial t}E_{\alpha\beta}(\mathbf{r},t) = \frac{\partial v_{p\alpha}}{\partial r_\mu}E_{\mu\beta} . \tag{3.16}$$

The free energy of the gel is then expressed as a functional of the tensor $\mathbf{E}(\mathbf{r},t)$. Following Onuki[17], I assume this to be given by

$$A = \int d\mathbf{r}\left[f(E_{\alpha\beta}(\mathbf{r},t))+ \frac{1}{2}K(\nabla\phi)^2\right] . \tag{3.17}$$

Hence $\dot{A}$ is calculated as

$$\begin{aligned} \dot{A} &= \int d\mathbf{r}\left[\frac{\partial f}{\partial E_{\alpha\beta}}\dot{E}_{\alpha\beta} + K(\nabla\phi)\cdot(\nabla\dot{\phi})\right] \\ &= \int d\mathbf{r}\left[\frac{\partial f}{\partial E_{\alpha\beta}}\cdot\frac{\partial v_{p\alpha}}{\partial r_\mu}E_{\mu\beta} -K(\nabla^2\phi)(\nabla\cdot\phi\mathbf{v}_p)\right] \\ &= -\int d\mathbf{r}\left\{\frac{\partial}{\partial r_\mu}\left(\frac{\partial f}{\partial E_{\alpha\beta}}E_{\mu\beta}\right) - K\phi\frac{\partial}{\partial r_\alpha}\nabla^2\phi\right\}v_{p\alpha} \end{aligned} \tag{3.18}$$

whence

$$F_{p\alpha}= \frac{\partial}{\partial r_\mu}\left(\frac{\partial f}{\partial E_{\alpha\beta}}E_{\mu\beta}\right) + K\phi\frac{\partial}{\partial r_\alpha}\nabla^2\phi = \frac{\partial}{\partial r_\beta}\sigma_{p\alpha\beta} \tag{3.19}$$

where

$$\sigma_{p\alpha\beta} = \frac{\partial f}{\partial E_{\alpha\beta}}E_{\beta\mu}- \sigma_{K\alpha\beta} \tag{3.20}$$

and $\boldsymbol{\sigma}_K$ is given by eq.(3.11). Thus eqs.(2.11) and (2.12) become

$$\mathbf{v}_p-\mathbf{v} = \frac{(1-\phi)^2}{\zeta}\nabla\cdot\boldsymbol{\sigma}_p \tag{3.21}$$

$$\eta_s\nabla^2\mathbf{v} - \nabla p + \nabla\cdot\boldsymbol{\sigma}_p =0 . \tag{3.22}$$

In the situation where the system is macroscopically at rest, eq.(3.21) becomes

$$\mathbf{v}_p = \frac{(1-\phi)^2}{\zeta}\nabla\cdot\boldsymbol{\sigma}_p . \tag{3.23}$$

To get a closed equation, we have to know $f(\mathbf{E})$ explicitly. An often employed form is

$$f(\mathbf{E}) = \frac{\phi}{2\phi_0}G_0E_{\alpha\beta}E_{\alpha\beta} + f_{mix}(\phi) . \tag{3.24}$$

The first term corresponds to the elastic energy of the polymer chain and the second term is the free energy of mixing. From eqs.(3.20) and (3.24), we have

$$\sigma_{p\alpha\beta} = \frac{\phi}{\phi_0} G_0 E_{\alpha\mu} E_{\beta\mu} - \Pi_{mix}\delta_{\alpha\beta} + \sigma_{K\alpha\beta} \tag{3.25}$$

where

$$\Pi_{mix} = \phi\frac{\partial}{\partial\phi} f_{mix}(\phi) - f_{mix}(\phi) \quad . \tag{3.26}$$

Eqs.(3.23) and (3.20) have been used in the problems of gel swelling [16,17,19].

Notice that the same stress tensor σ_p appears in the driving force of diffusion eq.(3.21) and mechanical balance (3.22). This is reasonable because the stress in a gel is transmitted through the polymer network. If there is a gradient in the stress, it will create a net force on the network, and will cause a motion of the network relative to the solvent.

3.3 Polymer solutions

We shall now consider polymer solutions. The distinction between polymer solutions and gels is that polymer solutions cannot sustain equilibrium shear stress. However this would not alter the essential features of the constitutive equation. We propose that **v** and $\mathbf{v}_p$ of the polymer solutions will satisfy the same equation as for gels i.e.,

$$\mathbf{v}_p - \mathbf{v} = \frac{(1-\phi)^2}{\zeta}\nabla\cdot\sigma_p \quad , \tag{3.27}$$

$$\nabla\cdot(\sigma_p - p\mathrm{I}) = 0 \quad , \tag{3.28}$$

where the solvent viscosity term in eq.(3.22) has been neglected since it is usually much smaller than the term of σ_p. Eq.(3.28) indicates that the macroscopic stress tensor is given by

$$\sigma = \sigma_p - p\mathrm{I} \quad . \tag{3.29}$$

Eqs.(2.2),(2.3),(3.27) and (3.28) give a complete description of the coupled phenomena of flow and diffusion in polymer solutions.

As in the case of gels, the same tensor σ_p appears in the diffusion equation and the mechanical balance equation. Again this is because the stress in the concentrated polymer solutions is transmitted by the entangled polymers: if there is a gradient in the stress, it will create a net force on polymer chains, and will cause the diffusion of the polymer relative to the solvent.

To get a closed equation, a constitutive equation for σ_p is needed. Since the deviatropic part of σ_p is equal to that of σ, i.e.,

$$\sigma_p - \frac{1}{3}I\ Tr(\sigma_p) = \sigma - \frac{1}{3}I\ Tr(\sigma) \ , \tag{3.30}$$

it can be determined by the measurement of mechanical quantities such as the shear stress or the normal stress difference. On the other hand, the isotropic part $Tr(\sigma_p)$ requires new information.

Little is known about the isotropic part of σ_p since conventional rheological measurements are done only for the deviatropic part. Here I will proceed by using a tentative model for the isotropic part.

At equilibrium, $Tr(\sigma_p)/3$ is equal to the osmotic pressure and depends on Φ only:

$$Tr(\sigma_p)/3 = -\Pi_o(\Phi) \ . \tag{3.31}$$

If the system is not in equilibrium, $Tr(\sigma_p)$ will generally be a functional of the deformation tensor (as can be seen in the case of gels). Here, for simplicity, we disregard the viscoelastic effect in the isotropic part, and simply assume eq.(3.31) for the nonequilibrium case. Thus

$$\sigma_p = \sigma^{(d)} - \Pi(\Phi)I \ , \tag{3.32}$$

$$\sigma = \sigma^{(d)} - pI \ , \tag{3.33}$$

where p is determined by the incompressible condition $\nabla\cdot\mathbf{v}=0$.
The deviatropic part $\sigma^{(d)}$ may be given by the conventional rheological constitutive equation. For example, if the stress is small, the linear viscoelastic model holds quite generally. In this case, σ_p may be written as

$$\sigma_p(t) = \int_{-\infty}^{t} dt' G(t-t') \left[\nabla\mathbf{v}_p + (\nabla\mathbf{v}_p)^{\dagger} - \frac{2}{3}I\nabla\cdot\mathbf{v}_p\right]_{\text{time } t'} - \Pi(t)I. \tag{3.34}$$

Notice that in the conventional rheological equation, $\nabla\mathbf{v}$ is used in the place of $\nabla\mathbf{v}_p$ in the above equation. If the diffusion of polymer is negligible, this does not matter. However, if $\mathbf{v}_p \neq \mathbf{v}$, one should use $\nabla\mathbf{v}_p$ instead of $\nabla\mathbf{v}$ since it is the deformation of polymers which is responsible for σ_p.

4. Applications

I will now discuss two problems using the above constitutive equation.

4.1 Swelling of polymer solutions:

As the first example, let us consider the swelling of a polymer solution in a tube (Fig.2). For simplicity we shall neglect the effect of the tube wall, and consider the one-dimensional diffusion along the tube axis only. Thus $\mathbf{v}$ and $\mathbf{v}_p$ depend on the coordinate x along the tube. The incompressible condition $\nabla\cdot\mathbf{v}=0$ and the boundary condition at the bottom of the tube leads to $v_x=0$. Thus eqs.(3.27) and (2.3) become:

$$v_{px} = \lambda \frac{\partial}{\partial x}(\sigma_{xx}^{(d)} - \Pi(\phi)) , \qquad (4.1\ )$$

$$\frac{\partial\phi}{\partial t} = - \frac{\partial}{\partial x}(\phi v_{px}) . \qquad (4.2\ )$$

For the linear viscoelastic model, we have

$$\sigma_{xx}^{(d)} = \frac{2}{3}\int_{-\infty}^{t} dt'\ G(t-t')\ \frac{\partial v_{px}}{\partial x} . \qquad (4.3\ )$$

Equations (4.1),(4.2) and (4.3) agree with those proposed by Brochard and de Gennes [5]. The same set of equations has been used to calculate the time correlation function of the concentration fluctuation.[4].

If one takes account of the effect of the tube wall, it will generally cause a macroscopic flow. Likewise, if one has a macroscopic particle suspended in the polymer solution, the particle will move as the diffusion of polymer takes place. Details can be analysed by use of the constitutive equation.

4.2 Concentration gradient in parallel plates

Next we shall consider the concentration profile in the shear flow between parallel disks shown in Fig.1. Let us take the cylindrical coordinate (r,θ,z). By symmetry, it can be shown that the stress and the concentration in the steady state depend on r only. From the incompressible condition $\nabla\cdot\mathbf{v}=0$, and the continuity equation $\nabla\cdot(\mathbf{v}_p\phi)=0$, we have $v_r=v_{pr}=0$. Thus the radial component of eq.(3.27) becomes

$$\frac{1}{r}\frac{\partial}{\partial r}\left[r(\sigma_{rr}^{(d)}-\Pi)\right] - \frac{\sigma_{\theta\theta}^{(d)}-\Pi}{r} = 0 , \qquad (4.4)$$

ie.,

$$\frac{\partial\Pi}{\partial\Phi}\cdot\frac{\partial\Phi}{\partial r} = \frac{\partial\sigma_{rr}^{(d)}}{\partial r} - \frac{\sigma_{rr}^{(d)}-\sigma_{\theta\theta}^{(d)}}{r} . \qquad (4.5)$$

The right hand side is a function of the shear rate $\dot{\gamma}(r)=\partial v_\theta/\partial z$, and may be expressed by the first and second normal stress differences N_1 and N_2:

$$N_1 = \sigma_{\theta\theta}^{(d)}-\sigma_{zz}^{(d)} \quad ; \quad N_2 = \sigma_{zz}^{(d)}-\sigma_{rr}^{(d)} . \qquad (4.6)$$

Since $\sigma_{rr}^{(d)}+\sigma_{\theta\theta}^{(d)}+\sigma_{zz}^{(d)}=0$, we have

$$\sigma_{rr}^{(d)} = -\frac{N_1+2N_2}{3} , \qquad (4.7)$$

$$\sigma_{\theta\theta}^{(d)} = \frac{2(N_1-N_2)}{3} . \qquad (4.8)$$

Thus

$$\frac{\partial\Pi}{\partial\Phi}\cdot\frac{\partial\Phi}{\partial r} = -\frac{\partial}{\partial\dot{\gamma}}\left(\frac{N_1+2N_2}{3}\right)\cdot\frac{\partial\dot{\gamma}}{\partial r} - \frac{N_1+N_2}{r} . \qquad (4.9)$$

Usually, N_2 is much smaller than N_1, and $\partial N_1/\partial\dot{\gamma} > 0$. Thus the right hand side of eq.(4.9) is negative. Since $\partial\Pi/\partial\Phi>0$, eq.(4.9) indicates that the concentration in the center is higher than the concentration in the rim. This result agrees qualitatively with the previous theory [1-3] based on the free energy argument. However, the physical explanation is essentially different. In the present theory, the concentration gradient is caused by the normal stress effect and has nothing to do with the free energy of the chain.

The distinction between the two theories becomes clearer for the cone-and-plate geometry, where the shear rate is uniform. According to the free energy argument, the concentration should be uniform in this geometry, while the present theory predicts that the concentration in the center should be higher than that near the edge: because the force which causes the Weissenberg effect essentially acts on polymers and causes the concentration gradient. Thus the two theories can be discriminated by experiments.

Crudely speaking, the concentration difference $\delta\Phi$ in the general inhomogeneous flow may be estimated by

$$\frac{\delta\Phi}{\Phi} \cong \frac{\delta N_1}{\Phi\left(\frac{\partial\Pi}{\partial\Phi}\right)} , \qquad (4.10)$$

where δN_1 is the typical magnitude of the spatial variation of the normal stress difference N_1. The concentration variation $\delta\phi$ can be a significant value for poor solvent in strong shear flow.

5. Conclusion

I have proposed a phenomenological constitutive equation that describes the coupling between the viscoelasticity and diffusion. This effect is important in various situations, but has been overlooked in conventional rheology. The constitutive equation I gave includes various hypotheses. An experimental check of the equation has fundamental importance.

Acknowledgement

This work emerged from a series of discussions with Prof. A. Onuki. I sincerely thank him for his stimulating input and valuable comments.

References

1 W. F. Busse, J. Polym. Sci. Part A 2 1261 (1967)

2 H. P. Schreiber and S. H. Storey, J. Polym. Sci. B 723 (1965)

3 A. B. Metzner, Y. Cohen and C. Rangel-Nafaile, J. NonNewtonian Fluid Mech. 5 449 (1979).

4 F. Brochard and P.G. de Gennes, Macromolecules 10 1157 (1977), F. Brochard, J. Physique 44 39 (1983)

5 F. Brochard and P. G. de Gennes, Physicochem. Hydr. 4 313-322 (1983).

6 N. L. Thomas and A. H. Windle, Polymer 23 529 (1982).

7 C. J. Durning, J. Polym. Sci. Phys. 23, 1831-1855 (1985), C. J. During and M. Tabor, Macromolecules 19 2220 (1986).

8 C. Rangel-Nafaile, A. B. Metzner and K. F. Wissbrun, Macromolecules 17 1187 (1984).

9 M. Tirrell, Fluid Phase Equilibria 30 367 (1986).

10 E. Helfand and G. H. Fredrickson, Phys. Rev. Lett. 62 2468 (1989).

11 A. Onuki, Phys. Rev. Lett. 62 2472 (1989)

12 R. B. Bird, O. Hassager, R. C. Armstrong and C. F. Curtiss, Dynamics of Polymeric Liquids (Wiley, N.Y. 1977).

13 K. Kawasaki and J. D. Gunton, in Progress in Liquid Physics ed. by C. Croxton (Wiley, N.Y. 1978). A. Onuki and K. Kawasaki, Ann. Phys. N.Y. 121 456 (1979).

14 H. Goldstein, Classical Mechanics, 2nd edt. (Addison Wesley, Reading, MA 1980).

15 M. Doi and S. F. Edwards, The theory of polymer dynamics (Oxford Univ. Press 1986).

16 T. Tanaka and D. J. Filmore, J. Chem. Phys. 70 1214 (1979)

17 A. Onuki, J. Phys. Soc. Jpn. 57 699 (1988).

18 A. Onuki, K. Yamazaki, and K. Kawasaki, Ann. Phys. (N.Y.) 131, 217 (1981); A. Onuki, Physica (Amsterdam) A140, 204 (1986).

19 K. Sekimoto and K. Kawasaki, J. Phys. Soc. Jpn. 56 2997 (1987), Physica (Amsterdam) A154, 384 (1989).

MC Study of Effects of Shear on the Phase Transition of Binary Mixtures

C.K. Chan

Institute of Physics, Academia Sinica,
Taipei, Taiwan 11529, Republic of China

It is well known that systems near critical points can be brought into a state far from equilibrium by a relatively small disturbance. A good example is a critical binary mixture under shear flow /1/. Experiments and theories have shown that the critical properties of the system can be drastically changed when critical fluctuations are exposed to shear. In 1979 Beysens et al./2/ found that critical fluctuations were strongly suppressed by a shear flow together with a lowered critical temperature T_c. Their findings were supported by the theories of Onuki et al. /3/. However, very recently, Cates et al. /5/ have shown that even when the concentration fluctuations are suppressed by the shear flow, the T_c of an isotropic-to-lamellar transition of a liquid crystal can still be raised. The main difference of the model of Onuki et al. and that of Cates et al. is that velocity fluctuations are neglected in the latter model. It is pointed out to the author by Onuki /5/ that it is the velocity fluctuations and its coupling to the composition fluctuations which lowers T_c.

Recently, we have performed a 2D Monte Carlo (MC) study of the effects of shear on a binary mixture /6/ where velocity fluctuations and its coupling to the composition fluctuations are neglected. The changes of the concentration of a binary mixture under shear at a point can be brought about by the random process of diffusion and the deterministic motion of the shear flow. We have assumed that the shear flow will not affect the thermal process and vice versa. Under these assumptions, the sole effect of shear flow is to set up a velocity gradient in the fluid and the thermal process will take place in the presence of this gradient.

The binary mixture is simulated on an N × N square lattice with two types of spins; $S = +1$ and $S = -1$. Shear flow on the lattice is simulated by moving the spins in the lattice in a fixed direction by different lattice points which equal to their distance from the top of the lattice. Under the action of shear, the first row of spins will be displaced by one lattice point and the second row by two lattice points and so on. Periodic boundary conditions are used in the direction of motion to make spins moved out of the lattice come back from the other end. This shear motion is applied to all the spins at the same time and it is called a shear step. The thermal interaction of the spins are given by the usual Ising-like Hamiltonian : $H = -J\sum_{i,j} S_i S_j$ where $\sum$ is the sum over nearest neighbours and J represents the strength of the interaction. The usual periodic boundary conditions are imposed only in the direction of flow. For the other direction, spins are pinned to the boundaries randomly. Equilibrium configurations at a particular temperature are generated by spin exchange dynamics. Spins pinned to the boundaries are not exchanged. The magnitude of the shear is controlled by changing the number of MC steps performed between two shear steps. The specific energy and total fluctuations are measured to determine the critical temperature.

In the simulation, it is found that critical fluctuations are strongly suppressed by the shear flow and the critical temperature T_c is raised as expected. There exists a limiting value for the shear higher than which the effect of shear becomes saturated. The phase separating domains are found to be in the form of parallel strips in the direction of flow. We have also estimated the critical exponent β to be 0.38 ± 0.05 which is closer to the mean field value. A higher T_c together with a greater β gives rise to the situation where the coexistence curve for a system under shear can cross that of the equilibrium one. The action of shear can lower or raise the phase separating temperature, T_{ps}, depending on the concentration of the system. This concentration dependence implies that there are at least two competing effects of the shear. One tries to raise T_{ps} and the other tries to lower it. It is interesting to note that the phenomenon observed in driven ionic conductions /7/ is very similar to those reported above. There is also a saturating electric field and the phase separating structure is also in the form of strips. It seems that both systems are reduced from 2D to quasi 1D systems. In fact, Kawasaki et. al. /8/ has predicted that the critical fluctuations of a binary mixture under shear (3D) will acquire 2D characters at the very high shear limit and there are also experimental observations /9/ which are consistent with the claim of Kawasaki et al. This dimensional reduction is possible in our model because spins on the same row (direction of flow) are not affected by the shear while interactions between spins on neighbouring rows are strongly suppressed by the flow.

This work is supported by the NSC of the Republic of China under the grant number NSC78-0208-M001-67.

REFERENCES

/1/. D. Beysens, M. Gbadamassi and B. Moncef-Bouarz, Phys. Rev. A, 28 , (1983), 2491 and references therein.
/2/. D. Beysens, M. Gbadamassi and L. Boyer, Phy. Rev. Lett., 43, (1979), 1253.
/3/. A. Onuki, K. Yamazaki and K. Kawasaki, Ann. Phys., 131, (1981), 217 and A. Onuki and K. Kawasaki, Supp. Prog. Ther. Phys. 64, (1978), 436.
/4/. M. E. Cates and S. T. Milner, Phy. Rev. Lett., 16, (1989), 1856.
/5/. A. Onuki, Private communications (1989)
/6/. C.K. Chan and L. Lin, Preprint (1989 Taiwan)
/7/. J.L Valles and J. Marro, J. Stat. Phys. 49, (1987), 89 and references therein.
/8/. T. Imeada and K. Kawasaki, Prog. Theor. Phys, 73, (1985), 559.
/9/. F. Perrot, C. K. Chan, D. Beysens, Europhys. Lett., 9, (1989), 65.

Domain Growth Under Shear Flow

T. Ohta[1], *H. Nozaki*[1], *and M. Doi*[2]

[1]Department of Physics, Ochanomizu University, Tokyo 112, Japan
[2]Department of Applied Physics, Nagoya University, Nagoya 464, Japan

Computer simulations of domain growth under a simple steady shear flow in a phase separating binary fluid have been performed in two dimensions to investigate domain deformation and the effect on macroscopic quantities such as the viscosity. The model equation is given in terms of the local concentration difference $C(\mathbf{r}, t)$ by

$$\partial_t C(\mathbf{r}, t) = L\nabla \{(1 - aC^2)\nabla(-D\nabla^2 C + F(C))\} - \mathbf{v}\nabla C \tag{1}$$

where $F(C) = -A \tanh C + C$. We have chosen the constants as L=1, D=0.5 and A=1.3 as in the study of spinodal decomposition [1] without the last term in (1). The external flow $\mathbf{v}$ is given in two dimensions by $\mathbf{v}(\mathbf{r})=(Sy, 0)$ with S the shear rate. We ignore the hydrodynamic interaction in the present simulations.

The cell dynamic version [1] of (1) is used for simulations. The system size is 111 x 111 with a suitable boundary condition. We have evaluated the anisotropic factor $Q_{xy}= -\langle(\partial_x C)(\partial_y C)\rangle$ which is proportional to the xy-component of the stress tensor. The result for several values of S is exhibited in Fig. 1 as a function of shear strain γ=St. We have found that the effective viscosity defined by $\Delta\eta = Q_{xy}/S$ obeys approximately a scaling law

$$\Delta\eta = S^{-\omega} f(\gamma) \tag{2}$$

where $\omega \approx 0.86$. This value is slightly larger than our theoretical estimate $\omega = 2/3$ for $a=0$ [2].

The pattern evolution of domains for the volume fraction 0.5 and $S= 1.2\times10^{-3}$ with a=0 in (1) is displayed in Fig. 2 where the black (white) region indicates C > (<) 0. Stretching, shear coagulation and recombination of domains can be clearly seen in Fig. 2.

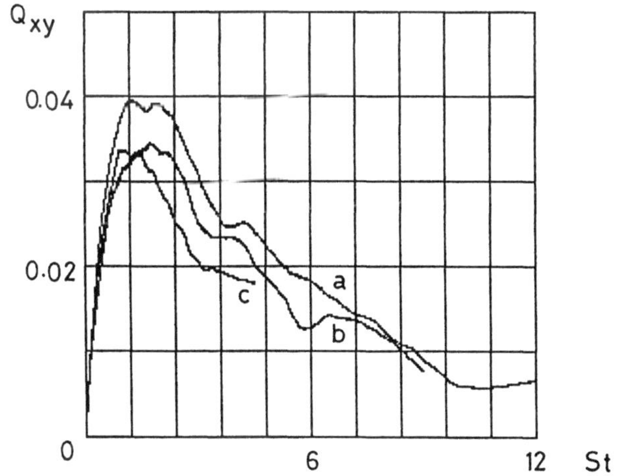

Fig. 1. Q_{xy} as a function of γ for S=0.0012 (a), 0.0006 (b) and 0.0003 (c).

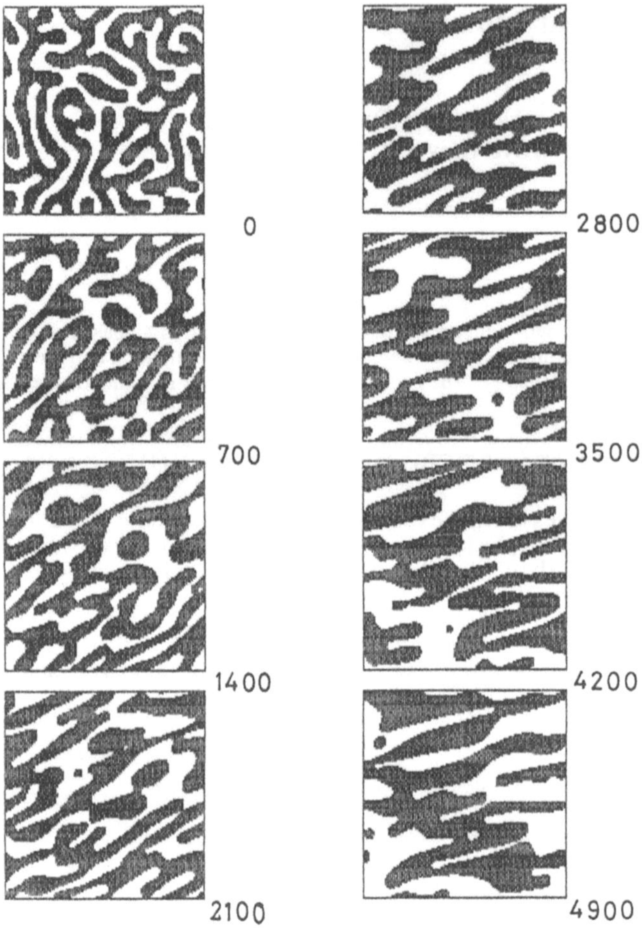

Fig. 2. Time-evolution of interconnected domains. The time is indicated below each figure.

In the meeting we presented a video movie [3] of simulations of a burst of a single domain as well as the time-evolution of interconnected domains as in Fig.1.

A part of the present results will be published in Ref. [2].

References

[1] Y. Oono and S. Puri, Phys. Rev. **A38**, 434 (1988).
[2] T. Ohta, H. Nozaki and M. Doi, submitted to Phys. Letters A.
[3] H. Nozaki, T. Ohta and R. Kobayashi, movie on "Domain Growth under Shear Flow".

Turbulent Flow of Dilute Polymer Solutions: An Approach to the Mechanism Producing Drag Reduction

A. Hatano

Department of Pure and Applied Sciences, College of Arts and Sciences, The University of Tokyo, Komaba 3-8-1, Meguro-ku, Tokyo 153, Japan

A self-consistent treatment of the turbulent flow of dilute polymer solutions is presented and the mechanism producing drag reduction is considered.

The theoretical features of turbulence in drag reducing polymer solutions are still unclear even though much work has been done. We present a formal framework of a self-consistent treatment of the turbulent flow of dilute polymer solutions.

In the form of Edwards-Freed theory [1], we can describe the coupled equation for the incompressible solvent velocity field $u(r,t)$, i.e., $\nabla\cdot u(r,t)=0$, and the coordinates of the polymer segments $\{R_i(s_i,t)\}$; $i=1,2,\ldots,N$, as follows.

$$\frac{Du(r,t)}{Dt} = -\frac{1}{\rho}\nabla p(r,t) + \nu\nabla^2 u(r,t) + \sum_i\int \sigma_i(t)\phi_i(r,t)ds_i + F(r,t), \quad (1)$$

$$\tau\frac{\partial}{\partial t}R_i(s_i,t) - \frac{3kT}{l^2}\frac{\partial^2}{\partial s_i^2}R_i(s_i,t) = -\sigma_i(s_i,t) + f_i(s_i,t), \quad (2)$$

where $D/Dt=(\partial/\partial t)+(u\cdot\nabla)$, ρ, p and ν denote density, pressure and kinematical viscosity of the solvent, and i, τ, l and s_i denote segment number, hopping time, segment length and contour length of the polymer, respectively. σ_i is the force between u and R_i, and F and f_i are external forces. $\phi_i(r,t)=\delta(r-R_i(s_i,t))$. Although turbulence is developed enough, we may assume that the flow is no-slip at the segment:

$$u(R_i,t) = \int\phi_i(r,t)u(r,t)dr = \dot{R}_i. \quad (3)$$

Eliminating p in Eq.(1) using $\nabla\cdot u=0$, we obtain a nonlinear equation of u as,

$$\frac{Du}{Dt} - \nu\Delta u - \int Q(r,r')\left(\nabla\frac{Du}{Dt}\right)_{r'}dr' = \int P(r,r')\{\sum_i\int\sigma_i(s_i,t)\phi_i(r',s_i;t)ds_i + F(r',t)\}dr', \quad (4)$$

where $P(r,r')=\delta(r-r')-Q(r,r')$; $Q(r,r')=(1/4\pi)\nabla_r\cdot|r-r'|^{-1}\cdot\nabla_{r'}$. But this equation cannot be solved exactly, an appropriate approximation is required.

We look for an approximation which would be suited for investigating the behavior of the turbulence of polymer solutions. For the purpose, a formal derivation is developed in this report. We decompose nonlinear terms of u in Eq.(4) into two parts: non-local damping term $h(r,r')$ and stochastic term $q(r,r')$, i.e.,

$$\int\{L_0\delta(r-r')+h(r,r')\}u(r')dr' = \int\{q(r,r')+P(r,r')[\sum_i\int\sigma_i\phi_i ds_i + F(r')]\}dr', \quad (5)$$

where $L_0=\partial/\partial t-\nu\Delta$. In the absence of polymer, we can reduce Eq.(5) to that of the direct interaction approximation for turbulence after some approximations [2].

Assuming the existence of the inverse operator $(L_0\delta(r-r')+h(r,r'))^{-1}\equiv L^{-1}(r,r')$ and using Eq.(2) and Eq.(5), Eq.(3) becomes

$$\dot{R}_i=\iiint \phi_i(r)L^{-1}(r,r')\{q(r',r'')+P(r',r'')[\sum_j(-\not{\Delta}_j R_j+f_j)\phi_j(r'')+F(r'')]\}drdr'dr'', \quad (6)$$

where $\not{\Delta}_i R_i \equiv \{\tau\partial/\partial t-(3kT/l^2)\partial^2/\partial s_i^2\}R_i$. By rewriting Eq.(6), the polymer motion under the turbulent flow is given by

$$\dot{R}_i+\sum_j\iiint \phi_i L^{-1}P\phi_j\not{\Delta}_j R_j drdr'dr''=\iiint \phi_i L^{-1}\{q+P[\sum_j\phi_j f_j+F]\}drdr'dr''. \quad (7)$$

We denote the resolvent L^{-1} by G, which is the turbulent flow propagator. Then,

$$G_{ij}\equiv\iiint \phi_i G(r,r')P(r',r'')\phi_j drdr'dr''=\zeta^{-1}\delta_{ij}+(1-\delta_{ij})G_{ij} \quad (8)$$

is the propagator of the turbulent velocity field between segments i and j, where ζ is the effective friction constant of the segment by the fluid. Effects of the term h, which causes the turbulence, spread to long distance. When h can be treated by perturbation, G can be expanded by the power series of h with $G_0=[L_0\delta(r-r')]^{-1}$ as

$$G_{ij}=\zeta^{-1}\delta_{ij}+(1-\delta_{ij})[G_0P-G_0PhG_0P+\cdots]_{ij}. \quad (9)$$

Since h, of course, is positive, it contributes to reducing the spring force of the polymer chain and thus the polymer will be stretched by the turbulence [3].

Now, we introduce a new operator K to $\{R_i\}$ as $K_{ij}\equiv\partial/\partial t\delta_{ij}+G_{ij}\not{\Delta}_j$, and assume the inverse of K_{ij} as $K_{ij}^{-1}K_{jk}=\delta_{ik}$. Then, R_i is formally given as

$$R_i=\sum_j\iiint K_{ij}^{-1}\phi_j(r)G(r,r')[q(r',r'')+P(r',r'')F(r'')]drdr'dr''+\sum_{jk}K_{ij}G_{jk}f_k. \quad (10)$$

Further, combining Eq.(2) with Eq.(10), we find

$$\sigma_i=-\not{\Delta}_i R_i+f_i=-\sum_j\iiint \not{\Delta}_i K_{ij}^{-1}\phi_j(r)G(r,r')[q(r',r'')+P(r',r'')F(r'')]drdr'dr''+\hat{f}_i, \quad (11)$$

where $\hat{f}_i$ is effective f_i. But, the terms of f_i are omitted because the sum of the external forces acting on the segments can be set to be zero.

Thus, the substitution of Eq.(11) into Eq.(5) yields our main equation,

$$\int[L_0\delta(r-r')+h]udr'=\int[q-P\sum_{ij}\iiint \phi_i\not{\Delta}_i K_{ij}^{-1}\phi_j G(q+PF)dr_1dr_2dr_3]dr'. \quad (12)$$

This equation tells us that as $\iint G(r',r_1)[q(r_1,r_2)+P(r_1,r_2)F(r_2)]dr_1dr_2$ can approximately be replaced by $u(r')$, the damping term h increases roughly as follows:

$$h(r,r') \to h(r,r')+\int\sum_{ij}P(r,r_1)\phi_i(r_1)\not{\Delta}_i K_{ij}^{-1}\phi_j(r')dr_1, \quad (13)$$

i.e., the effective damping increases. The last term represents the effect of polymer. Therefore, we can see that polymer reduces the drag.

We have formally presented a formulation of turbulent flow with polymer solutions. We plan to come to the quantitative analysis based on our formulation.

References

1. S. F. Edwards and K. F. Freed, J. Chem. Phys. 61, 1189 (1972).
2. R. H. Kraichnan, Phys. Fluids 9, 1728 (1966); J. R. Herring and R. H. Kraichnan, Lect. Notes Phys. 12, 148 (1972).
3. R. Armstrong and M. S. Jhon, J. Chem. Phys. 79, 3143 (1983).

Anomalous Phase Separation Behavior in a Binary Mixture of Poly(vinyl methyl ether) and Water Under Deep Quench Conditions

H. Tanaka[1] *and T. Nishi*[2]

[1]Department of Applied Physics and Applied Mechanics, Institute of Industrial Science, University of Tokyo, Minato-ku, Tokyo 106, Japan
[2]Department of Applied Physics, Faculty of Engineering, University of Tokyo, Bunkyo-ku, Tokyo 113, Japan

1. Introduction

Generally, the dynamics and morphology of phase separation strongly depends on the quenching conditions[1]. From this standpoint, phase separation phenomena are generally divided into nucleation-growth (NG) and spinodal decomposition (SD). In the unstable region, decomposition (a coarse graining process) usually proceeds faster with an increase in the quenching depth because the thermodynamic driving force for phase separation becomes larger. Here we report an anomalous, local phase-separation behavior which seems not to belong to the conventional categories described above. This phase separation is unusual because it does not accompany a coarse graining process after the formation of droplets. It has been observed in a binary mixture of poly(vinyl methyl ether)(PVME) and water at deep-level quenching conditions near the critical composition[2]. A digital image analysis (DIA)[3-7] has been used for the quantitative study.

2. Results and Discussions

Figure 1 shows the phase diagram of this system. In the shaded area (at the point expressed by the black circle), an anomalous phase separation was observed. At the point expressed by the open triangle, it was difficult to distinguish whether phase separation is usual or not. In the shaded area, small droplets of PVME are formed immediately after the deep quench. The droplets move around rapidly by Brownian motion. The situation is schematically drawn in Fig.2. The droplets seem to be bound in some sort of potential. The system is almost in a stationary state and phase separation does not proceed further for a very long time. The droplets are independent of each other and do not grow by a coalescence mechanism. It should also be mentioned that the morphological transition between the usual SD and unusual local phase separation exactly coincides with the dynamic transition of the motional state of the droplets. The droplets move vigorously above the transition, while the motion of droplets slows down below the transition, probably because of growth or coalescence of droplets. In this system, a typical SD-type phase separation with coarse graining takes place at the shallow quenching condition (below the dotted line in

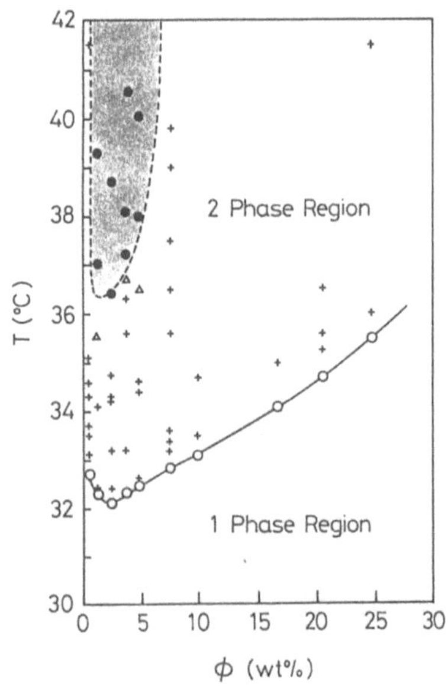

Fig.1 The phase diagram of the system. The solid line is a binodal line. In the shaded area,the anomalous phase separation is observed. At the point indicated by the crosses, a typical SD-type phase separation is observed.

Fig.2 Schematic figure for anomalous phase-separation behavior. Droplets do not grow by coalescence. They move vigorously by Brownian motion.

Fig.1)[7], while the anomalous behavior is observed at a deep quenching condition near the critical composition.

There are two possible physical mechanisms (energetic and dynamic) explaining this unusual phase separation: (1) Steric repulsions between droplets. (2) The competition between the contact time of the droplets during collision (τ_C) and the characteristic time required for coalescence by the interdiffusion of polymers between droplets (τ_D). In this case, the droplet structure is stabilized by a dynamic factor, and not thermodynamically. This might be a new type of phase in condensed matters, which is not stabilized energetically. Theoretical understanding of the phenomenon involves further study. The details regarding the anomalous phase separation will be published elsewhere.

References

1) See for a review, J.D.Gunton, M.San Muguel, and P.Sahni, in Phase Transition and Critical Phenomena, ed. C.Domb and J.H.Lebowitz (Academic, London, 1983), Vol.8.
2) H.Tanaka and T.Nishi, Jpn.J.Appl.Phys. **27**, L1787 (1988).
3) H.Tanaka, T.Hayashi, and T.Nishi, J.Appl.Phys. **59**, 653 (1986).
4) H.Tanaka, T.Hayashi, and T.Nishi, J.Appl.Phys. **59**, 3627 (1986).
5) H.Tanaka and T.Nishi, Phys.Rev.Letters **59**, 692 (1987).
6) H.Tanaka and T.Nishi, Jpn.J.Appl.Phys. **27**, L1783 (1988).
7) H.Tanaka, T.Hayashi, and T.Nishi, J.Appl.Phys. **65**, 4480 (1989).

Study of Pattern Formation Dynamics in a Polymer Mixture Undergoing Phase Separation by Digital Image Analysis

H. Tanaka[1], *T. Hayashi*[2], *and T. Nishi*[2]

[1]Department of Applied Physics and Applied Mechanics, Institute of Industrial Science, University of Tokyo, Minato-ku, Toyko 106, Japan
[2]Department of Applied Physics, Faculty of Engineering, University of Tokyo, Bunkyo-ku, Tokyo 113, Japan

1. Introduction

We have applied a digital image analysis (DIA) method to the study of the coarse-graining process[1] at the late stage of phase separation in polymer mixtures[2-9]. At the late stage, the sharp interface between two phases is formed and the motion of the interface is important. The conventional scattering technique is not suitable for analyzing the phase-separated pattern at the late stage since the scattering measurement in a low k range is difficult because of the effect of a direct beam. Only the real-space operations in DIA make it possible to extract the geometrical characteristics from the pattern. We will discuss the characteristics of pattern formation at the late stage of phase separation in polymer mixtures.

2. Experimental

The samples studied were a mixture of polystyrene (PS) and poly(vinyl methyl ether) (PVME) which has a LCST (lower critical solution temperature)-type phase diagram. The phase-separated structures were directly observed under phase contrast microscopy. The pattern was digitized by an image processor and the various numerical operations were performed for a digitized image.

3. Results and Discussions

The phase-separated structures can be grouped into the following two types: (1)The interconnected structure. (2) The droplet-like pattern. The interconnected structure can be characterized by the structure factor S(k) and the shape and length of interface; and the droplet-like pattern by S(k), the number, total area, and shape of droplets. The real-space operations in DIA are important for characterizing the geometrical characteristics of droplet patterns. In both cases, the concentration distribution can be calculated by a simple operation in DIA and it can not be obtained by other methods. From the temporal change of the total area of droplets, it becomes clear that the concentration change finishes at the rather earlier stage and in the late stage only coarse-graining proceeds without changing concentration. In this regime, the black&white operation is suitable for the analysis of pattern. The droplets coagulate with each other by direct collision. It is revealed that the size of the droplets changes in an intermittent manner. The morphology of the phase-separated pattern changes to

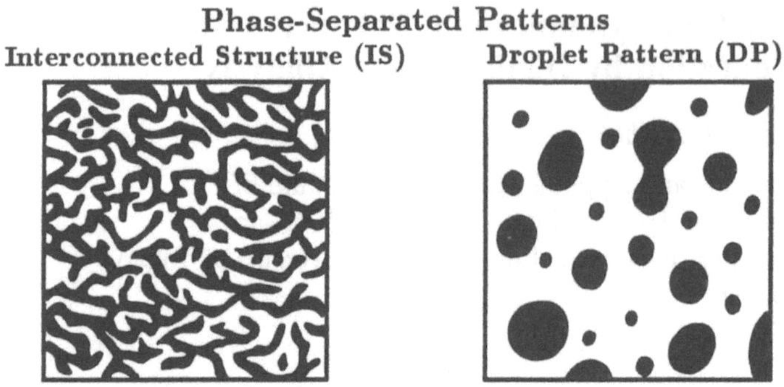

DIA Operations

k-Space Operations

o 2D Power Spectrum of an Original Image — S(k)

Real-Space Operations

o Total Interface Length	(IS, DP)
o Curvature Distribution	(IS, DP)
o Motion of Interface	(IS, DP)
o Motion of Droplets	(DP)
o Point Pattern Analysis	(DP)
o Number of Droplets	(DP)
o Shape of Droplet	(DP)
o Distribution of Droplet Size	(DP)

Fig.1 DIA operations useful for the study of phase-separated structures.

minimize the interfacial energy of the system. The temporal change in the number of droplets and the total interfacial length is found to be described by simple power laws[8]. The motion of interface and that of the center of mass of droplets has been successfully studied by a dynamic image analysis[9]. Quantitative and theoretical analysis will be shown elsewhere.

References

1) See for a review, J.D.Gunton, M.San Muguel, and P.Sahni, in Phase Transition and Critical Phenomena, ed. C.Domb and J.H.Lebowitz (Academic, London, 1983), Vol.8.
2) H.Tanaka, T.Hayashi, and T.Nishi, J.Appl.Phys. **59**, 653 (1986).
3) H.Tanaka, T.Hayashi, and T.Nishi, J.Appl.Phys. **59**, 3627 (1986).
4) H.Tanaka and T.Nishi, Phys.Rev.Letters **59**, 692 (1987).
5) H.Tanaka and T.Nishi, Jpn.J.Appl.Phys. **27**, L1783 (1988).
6) H.Tanaka and T.Nishi, Jpn.J.Appl.Phys. **27**, L1787 (1988).
7) H.Tanaka, T.Hayashi, and T.Nishi, J.Appl.Phys. **65**, 4480 (1989).
8) H.Tanaka, K.Takashima, T.Hayashi, and T.Nishi, Polym.Prepr.Jpn. **38**, 3365 (1989).
9) H.Tanaka, T.Hayashi, and T.Nishi, Polym.Prepr.Jpn. **38**, 4227 (1989).

Dependence of Molecular Weight and Tacticity of Poly(methylmethacrylate) (PMMA) on Crystallization of Poly(ethyleneoxide) (PEO) in PEO/PMMA Blends

M. Takahashi, N. Harasawa, and H. Yoshida

Department of Industrial Chemistry, Tokyo Metropolitan University, Fukazawa, Setagaya-ku, Tokyo 158, Japan

Abstract. Crystallization of poly(ethyleneoxide) (PEO) in blends with poly(methylmethacrylate) (PMMA) was investigated by a differential scanning calorimeter (DSC). The heat of melting, ΔHm, and the activation energy for crystallization, Ea, depended on PMMA tacticity. The crystallization rate became slower with increasing molecular weight of PMMA.

1. Introduction

Crystallization of crystalline/amorphous polymer blends is influenced by the physical properties of the amorphous component, such as molecular weight, tacticity, glass transition temperature, etc. This study aims at investigating the dependence of molecular weight and tacticity of PMMA on crystallization of PEO in PEO/PMMA blends.

2. Experimental

PEO (nominal molecular weight = 50,000 ± 10,000) used was the product of Wako pure chemical industries Ltd. Two kinds of atactic PMMA: a-PMMA1 (Mw = 7.7×10^5, Mw/Mn = 2.1) supplied by Mitsubishi Rayon, and a-PMMA2 (nominal molecular weight = 63,900, Mw = 5.5×10^4, Mw/Mn = 1.1) supplied by Pressure Chemical Co. were used. Isotactic PMMA (i-PMMA, ^{1}H-NMR triad [I] = 0.93, Mw = 2.3×10^4, Mw/Mn = 1.4) and syndiotactic PMMA (s-PMMA, ^{1}H-NMR triad [S] = 0.60, Mw = 2.0×10^5, Mw/Mn = 2.1) were synthesized by anionic polymerization. The values of Mw and Mw/Mn were determined by GPC using polystyrene as the standard. Blend samples were obtained by casting from chloroform solution. The isothermal crystallization process was measured by DSC (Seiko DSC 200 and Perkin Elmer DSC1B). Melting temperature (Tm), glass transition temperature (Tg) and heat of melting (ΔHm) were evaluated by heating at 10 K/min.

3. Results and Discussion

The crystallization of blend samples was observed at the PMMA weight fraction (W_{PMMA}) below 0.6. The blends with W_{PMMA} above 0.2 will never be homogeneous even if the temperature is raised after the nucleation occurs. Blend samples showed a broader melting peak than that of PEO. This may reflect the broadness of size distribution of the crystal domain. From the total area of melting peak, we evaluated the values of ΔHm which are listed in Table 1. The value of ΔHm shows scatter. However, it seems to depend on W_{PMMA} and PMMA tacticity independently of PMMA molecular weight.

As for the a-PMMA, the rate of crystallization becomes slower with increasing molecular weight. The equilibrium melting temperatures (Tm^0), which were determined by the Hoffman-Weeks plot, decreased with increasing W_{PMMA}. This means that the interaction parameter,

χ, of PEO/a-PMMA is negative [1],[2],[3]. Tm^0 were identical for the blends with the same W_{PMMA} within experimental error. Therefore, the difference of the rate of crystallization between both blends seems to be due to the effect of PMMA molecular weight on the co-operative diffusion. The activation energy of crystallization Ea was obtained from the Arrhenius plots of the half time of crystallization, which was the time for the crystallinity to become 50%. The obtained Ea were listed in Table 1. In the measured W_{PMMA}-region from 0.05 to 0.2, the obtained values of Ea depended on PMMA tacticity independently of molecular weight and composition.

Table 1. The values of ΔHm and Ea for PEO and PEO/PMMA blends

Sample	W_{PMMA}	ΔHm (J/g)	Ea (kJ/mol)
PEO	—	186	660
PEO/i-PMMA	0.053	161	577
	0.144	158	
	0.151	149	
	0.201	140	
PEO/s-PMMA	0.052	164	457
	0.101	154	
	0.152	141	
	0.202	132	
PEO/a-PMMA1	0.051	147	429
	0.055	149	
	0.101	137	
	0.103	116	
	0.155	113	
	0.194	135	
	0.200	120	
PEO/a-PMMA2	0.060	146	409
	0.108	150	
	0.151	155	
	0.187	116	
	0.20	119	

References

1. E. Martuscelli, C. Silvestre, M. L. Addonizio, and L. Amelino, Makromol. Chem. **187**, 1557 (1986).
2. H. Ito, T. P. Russel, and G. D. Wignall, Macromolecules **20**, 2213 (1987)
3. T. P. Russel, H. Ito, and G. D. Wignall, Macromolecules **21**, 1703 (1988)

Concentration-Dependent Double-Step Transition of the Diffusion Coefficient in Dilute Solutions of Styrene–Butadiene Diblock Copolymer in Selective Solvent

Y. Tsunashima

Institute of Chemical Research, Kyoto University, Uji, Kyoto-fu 611, Japan

Abstract. A poly(styrene)-block-poly(butadiene) copolymer dissolved in n-decane exibits a double-step transition in the diffusion coefficient D(c) at the extremely dilute polymer concentration ranging from 1×10^{-6} to 7.24×10^{-3} gcm^{-3} at 25 °C. The copolymer disperses monomolecularly only in the extremely dilute solution region limited to $c<c_1$ with $c_1 = 3.8\times10^{-6}$ gcm^{-3}. At $c>c_1$ (first onset concentration), micellization of some kind occurs due to the strong attractive interaction between polystyrene (PS) subchains, the copolymers associating intermolecularly. The micelle is huge and nearly monodisperse in size and is constructed from loose packing of PS subchains. At $c>c_2$ (second onset concentration with $c_2 = 1.1\times10^{-4}$ gcm^{-3}), the PS subchains collapse into a hard and compact core, while the polybutadiene (PB) subchains swell well and cover the core, forming approximately a "two-phase" concentric sphere. This micelle exibits an internal motion which may be caused by the concentration fluctuation of the highly extended PB chains relative to the hard core of PS chains.

1. Introduction

When AB type diblock copolymers dissolve in their selective solvents, i.e., solvents that are good solvents to one part B of the copolymer subchains but are precipitants to the other part A, the insoluble part A has to be protected in some ways with the help of the subchain B. Microphase separation on the single- or multi-molecular scale is a typical way. The former type of microphase separation may occur in very dilute solution [1] and it realizes intramolecular segregation. In this case, a strong attractive interaction takes place between pairs of segments of the subchain A in the same molecule. The single molecule thus constructs a structure which consists of the core of part A surrounded by the swollen part B. The latter type of microphase separation may be realized at slightly higher polymer concentration than the case of the former. The attractive interactions become dominant between the segment pairs of part A in different molecules. Association of a large number of single molecules results in micellar formation of various forms, depending on the molecular architecture, the solution concentration, and so on [2].

The details of the microphase separation have been little examined in solution, especially when the polymer concentration is limited to the very dilute region as mentioned above. The main reason is the lack of a method sensitive enough to detect low signals coming from such a solution system. However, dynamic light scattering spectroscopy gives useful information on this system if the spectrometer is in an excellent grade in its coherence condition [3]. This paper presents the first and exciting results obtained for the diblock copolymer solutions by using the dynamic light scattering method [4].

2. Experiments and Results

The PS-PB diblock copolymer was prepared by living polymerization, the weight average molecular weight M_w being 9.73×10^4 ($M_w/M_n = 1.03$) and the PS content 29.3 wt%. Dynamic light scattering measurements with our laboratory-made 512 channel time-interval correlator gave precisely normalized autocorrelation functions $A(\tau)$ of the light intensity scattered from the solution at scattering angles ranging

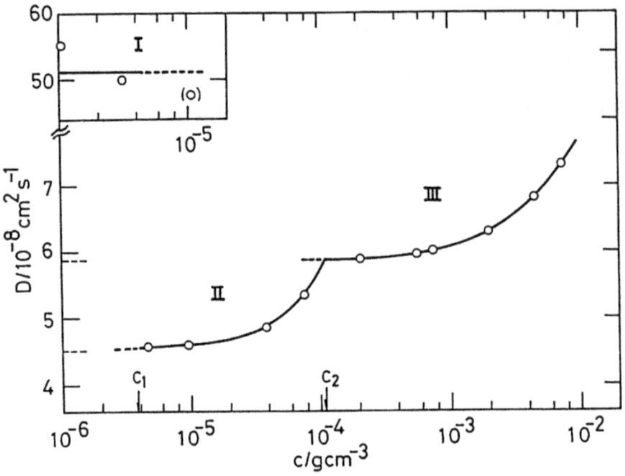

Figure 1. Concentration dependence of the translational diffusion coefficient D(c) measured for polystyrene-block-polybutadiene copolymer in n-decane at 25 C.

from 10° to 150°. A Spectra-Physics 3W argon ion laser equipped with an etalon and fine optical alignments assured a large coherence area and a high frequency of A(τ).

Figure 1 illustrates results of D(c) as a function of the polymer concentration c. In the extremely dilute solution region of $c<3.8\times10^{-6}gcm^{-3}$ (Region I), A(τ) showed single-exponential decays and the diffusion coefficient extrapolated to zero concentration, D_0, was $51\times10^{-8}cm^2s^{-1}$, which gave the Stokes radius $R_H \equiv k_BT/6\pi\eta_0D_0$ = 5.0 nm. This R_H value is about one third of that of the whole copolymer in good solvent (benzene) but is very close to 4.5 nm which was the value estimated for PS of exactly the same M_w as that of the PS subchain in the copolymer under the assumption that the PS was dissolved in benzene. The copolymer thus disperses molecularly in Region I. In the single molecule, the PB part swells well and protects the collapsed PS part, i.e., microphase segregation occurs intramolecularly.

In the highest concentration region studied, where $c>1.1\times10^{-4}gcm^{-3}$ (Region III), the first cumulant Γ_e changed its scattering vector (q) dependences from nearly constant to negative as c increased. Its simple extrapolation to c=0 and q=0 gave that $D_{0,app} = 5.87\times10^{-8}cm^2s^{-1}$ (R_H = 43.5 nm) and the concentration coefficient $k_D = 34.8cm^3g^{-1}$. The R_H value is about ten times larger than that in Region I. The molecular weight of this particle, 9.7×10^6, which was obtained from viscosity measurement in Region III, indicated that ca. 100 pieces of copolymers associate intermolecularly and form a micelle with a compact core of PS chains surrounded by swollen PB chains. Analysis based on a "two-phase" concentric spheres model revealed that the core radius is 10.2 nm, the radius of gyration (R_G) of PB chains in the shell is 19.7 nm, and the R_G of the micelle as a whole is 33.8 nm. The R_G value for PB indicates that the PB chains stretch highly in the shell region. On the other hand, the negative q dependence of Γ_e at large c is related to the appearance of a new mode characteristic of copolymer chain nature. The mode is due to the relative center-of-mass motion of PB chains with respect to the PS core.

In the intermediate concentration region defined by $3.8\times10^{-6}<c<1.1\times10^{-4}gcm^{-3}$ (Region II), A(τ) showed again single-exponential decays and the simple extrapolation of D(c) to c=0 gave that $D_{0,app} = 4.49\times10^{-8}cm^2s^{-1}$ (R_H = 56.8 nm) and k_D = 2450 cm^3g^{-1}. In this region, the copolymers form some kind of micelles which are of huge size but are nearly monodisperse. The micellar structure is still unknown because this structure requires a large positive k_D value as described above. Cylindrical rods and large hollow spheres cannot give such a k_D value.

3. References

1. C. Sadron, Pure and Appl. Chem. **4**, 347(1962)
2. Z. Tuzar and P. Kratochvil, Adv. Colloid Interface Sci. **6**, 201(1976).
3. Y. Tsunashima, M. Hirata, and Y. Kawamata, Macromolecules **23**, 1089(1990).
4. Y. Tsunashima, Macromolecules **23**, 2963(1990).

Dynamic Electric Multipole Model for Colloid Particles

S. Yoshino

Department of Physics, Faculty of Science, Nagoya University,
Chikusa-ku, Nagoya, Japan

Abstract. Colloid particles form many pairs with separations of several or tens of micrometers between the particles in a pair, which is two or three orders of magnitude larger than the value calculated by DLVO theory. To explain the experimental results, I propose the dynamic electric multipole model, which accounts for the effect of deformation of the ion atmosphere caused by the movement of the colloid particles.

1. Introduction

The movement and interaction of fine particles in a solvent have been investigated in three realms: the Brownian movement in physics, the mutual interaction of colloids in chemistry, and the interaction between bio-colloids in biology. Since the model of Brownian particles has a hard core potential and no charge, the particles move independently of each other. The model of colloid particles has a repulsive force caused by an electric double layer, and the attractive force arises from the van der Waals force on the colloid surface. The length of the secondary minimum of the total potential is a maximum of 20 nm [1]. In biology, the mutual interaction of bio-colloids is regarded as a short-range force under the physiological salt concentration. Therefore almost all scientists regard fine particles in solution as having only short-range interactions.

However, from my investigations in biophysics, I began to doubt whether the mutual interactions of fine particles in solution are only short range. I planned an experiment to reveal the long-range interactions between fine particles in solution. From observations of the distribution and movement of the particles through an optical microscope, I found many new unexplainable results [2], including ultra-long-range interactions. I have proposed a model for the ultra-long-range interaction, which was reported preliminarily in my previous paper [3].

2. Example of Experimental Results

The movement and distribution of the particles do not show an independent nature, but show a characteristic appearance. Much pair formation often occurs in solution. The probability of pair creation depends on the salt concentration. There are two typical separations of a pair: several micrometers and tens of micrometers. The lifetime of a pair is long, e.g. 1 hour. The lifetime depends on the concentration of the colloid, because a particle of a pair can change partner when a particle of a pair comes close to a particle of another pair.

3. Dynamic Electric Multipole Model

DLVO theory hardly explains the above experimental results [1]. I have investigated which assumptions of the theory cause discrepancies with real systems. I noticed two fundamental assumptions: the fixation of colloid particles in space, and the lack of distortion of the ion atmosphere. I proposed the following model for the colloidal interaction, which assumes distortion of the ion atmosphere caused by the movement of the colloid itself.

A colloid particle in solvent moves all the time by translational and rotational Brownian motion, and the movement of the colloid causes solvent flow around the particle. Since the ions in the ion atmosphere move with the flow, the collective motion of the ions causes a deformation of the ion atmosphere. The interaction between the deformed ion atmosphere and the surface charge on the colloid causes the ion atmosphere to return to its original position. However, in the process, a time lag is introduced into the time relation between the driving force and the displacement. The deformed charge distribution of the ion atmosphere and the core colloid forms an electric multipole.

The deformation of the ion atmosphere can be described as follows. We consider a spherical colloid, and use the Debye sphere for the form of the ion atmosphere. If the radius of the Debye sphere in DLVO theory is R_0, the deformation of the Debye sphere is given by

$$R_{(\theta,\phi,\tau,t)} = R_0(1 + \sum_{\ell} \sum_{m} \alpha_{\ell m(\tau,t)} Y^{*}_{\ell m(\theta,\phi)}) \ , \tag{1}$$

where $Y_{\ell m(\theta,\phi)}$ are the spherical harmonics and $\alpha_{\ell m(\tau,t)}$ are the deformation parameters which determine the deformed Debye sphere shape. The subscript m takes values from $-\ell$ to $+\ell$, that is, there are $2\ell+1$ modes of deformation of order ℓ. The first mode of deformation corresponds to a dipole mode ($\ell=1$). The second mode corresponds to a quadrupole mode ($\ell=2$). Each mode with m=0 in all ℓ values has symmetry with respect to an arbitrary rotation about the z-axis and therefore it represents an axially symmetric deformation of the Debye sphere. τ is the relaxation time of the deformation, which ranges widely from about a millisecond to a second. Such a deformation of the Debye sphere corresponds to the deformation of the electric charge distribution, that is, the $m\ell$-th deformation correspond to the $m\ell$-th multipole. Thus the dynamic multipole $M_{\ell m(\tau,t)}$ is written

$$M_{\ell m(\tau,t)} = \int \alpha_{\ell m(\tau,t)} \rho_{(\mathbf{r})} r^{\ell} Y_{\ell m} d\mathbf{r}, \tag{2}$$

where $\rho_{(\mathbf{r})}$ is the charge density in the ion atmosphere and the central colloid particle. This is the origin of the ultra-long-range attractive and repulsive interactions.

The lifetime of each multipole is of the order of the Brownian movement. However, the lifetime of a pair is not so short. The origin of the long lifetime may be a resonance between the multipoles.

From the model, it may be predicted that the shear flow creates a quadrupole whose degree depends on the magnitude of the shear flow, and the quadrupole-quadrupole interaction makes a pair of fine particles. The experimental results show many pairs, and the separation of the pairs depends on the degree of shear flow. These results support this model.

References

[1] Verwey E J W and Overbeek J Th G: Theory of the Stability of Lyophobic Colloids (Elsevier,Amsterdam 1948)

[2] Yoshino S and Imai N: R.P.P.P.J. **28**, 41 (1985)

[3] Yoshino S: in Ordering and Organization in Ionic Solutions ed.by Ise N and Sogami I (World Scientific,Singapore 1987)p.449

Part V

Phase Transitions in Two Dimensions

Experimental Studies of Phase Transitions and Pattern Formation in Two Dimensions

C.M. Knobler, K. Stine, and B.G. Moore

Department of Chemistry and Biochemistry, University of California, Los Angeles, CA 90024, USA

Abstract. The technique of fluorescence microscopy has been utilized in studies of the equilibrium phase behavior and the kinetics of phase transitions and pattern formation in monolayers at the air/water interface. A brief description of the equilibrium phase diagram of fatty acid monolayers is presented and quantitative studies of the evolution of two-dimensional gas-liquid foam structures in such systems are reported. Shape instabilities observed in monolayers of esters are described qualitatively.

1. Introduction

During the past decade there has been remarkable progress in our understanding of the dynamics of phase transitions, the result of advances in experiment, theory, and computer simulation. A key parameter, easily investigated in theory and simulation, is the dimensionality of the system. Experimenters have more difficulty in switching dimensions, however, and with few exceptions the experiments have been carried out in three-dimensional systems.

The problem of carrying out experimental studies of phase transitions in two dimensions is two-fold. One must find systems for which experiments are practical and one must then determine the equilibrium phase behavior in order to be able to define the conditions under which dynamic measurements are being carried out and the paths being followed. In this paper we will describe some recent attempts to carry out such an experimental program. The two-dimensional systems studied are monolayers of amphiphilic molecules at the air/water interface.

Monolayers have long been considered as the archetypes of experimentally realizable two-dimensional systems, but kinetic studies on them have become practical only with the introduction of techniques for the direct visualization of the monolayer phases, chief among them fluorescence microscopy. The imaging methods have also uncovered a dazzling array of patterns that form when systems are driven at phase transitions; some patterns may also be associated with equilibrium states.

We will begin in Section 2 with a brief discussion of monolayers, the variables with which they can be described, and the methods that can be employed in their study. In Section 3 we will review recent investigations of the equilibrium phase diagram of pentadecanoic acid, which is typical of that of many amphiphiles. A quantitative study of the growth of two-dimensional foams is described in Section 4. The paper concludes with more qualitative studies of instabilities and patterns.

Springer Proceedings in Physics, Vol. 52 **Dynamics and Patterns in Complex Fluids**
Editors: A. Onuki · K. Kawasaki

2. Langmuir Monolayers: Background and Experimental Methods

Our focus will be on *Langmuir* monolayers, which are monomolecular films at an air/water interface, in contradistinction to *Langmuir-Blodget* films, in which the substrate is the surface of a solid. The substances that can be spread as monolayers are amphiphiles, molecules that contain both polar (hydrophylic) and nonpolar (hydrophobic) moieties. Long-chain fatty acids are classic examples of amphiphiles that form monolayers. In these molecules, which have the general formula RCOOH, the carboxylic acid "head group" -COOH is embedded in the water substrate and R, the hydrocarbon "tail", remains in the air, Fig. 1. The tail must be sufficiently long that the acid has a very low solubility, which for practical purposes sets a lower limit at tetradecanoic acid, for which $R = C_{13}H_{27}$–. Other common amphiphiles are alcohols, esters, and phospholipids [1,2,3], Fig. 1.

FATTY ACIDS

$CH_3(CH_2)_{13}COOH$ pentadecanoic acid

$CH_3(CH_2)_{16}COOH$ octadecanoic acid (stearic acid)

ESTERS

$CH_3(CH_2)_{13}COOCH_3$ methyl pentadecanoate

$CH_3(CH_2)_{16}COOC_2H_5$ ethyl octadecanoate

PHOSPHOLIPIDS

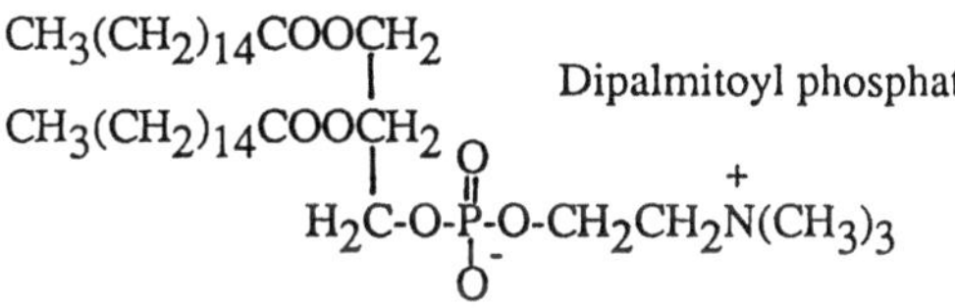

Dipalmitoyl phosphatidylcholine (DPPC)

Fig. 1. Typical monolayer-forming amphiphiles.

Monolayers are usually prepared by depositing dilute solutions of the amphiphile in a volatile solvent onto a clean water surface; the monolayer spreads when the solvent evaporates. The thermodynamic state of the monolayer is determined by the two-dimensional density (or area per molecule, A), the temperature T, and the surface pressure π; the average density of the monolayer either by making additions to the surface at constant area or by changing the area available to the film by sliding a barrier along the surface of the water. Surface pressures can be determined by use of the relation

$$\pi = \gamma_0 - \gamma$$

where γ_0 is the surface tension of pure water and γ is that when the surface is covered by the monolayer.

The equilibrium phase diagrams of monolayers have generally been determined from π,A isotherm measurements; a series of such isotherms for pentadecanoic acid is shown in Fig. 2. (The subphase is water at pH 2 in order to suppress the ionization of the carboxyl head group.) As in p,V isotherms for pure three-dimensional systems, horizontal slopes are associated with two-phase coexistence regions. One such region is evident at high areas and low pressures; it is ascribed to an equilibrium between a two-dimensional gaseous phase and a condensed phase that is generally called the *liquid-expanded* (LE) phase.

The kink in the isotherms at higher pressures is taken as the onset of another phase transition, that between the LE phase and another condensed phase that has been called the

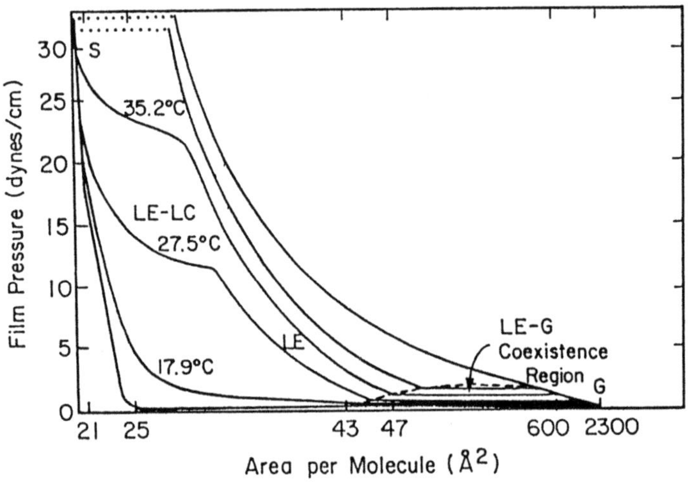

Fig. 2. Surface pressure - area isotherms. The isotherms shown are schematic but the temperatures indicated are those that apply to isotherms of pentadecanoic acid.

liquid-condensed (LC) phase. Although the isotherms shown here do not have a horizontal slope in this region, flat isotherms have been observed at the LE-LC transition in more recent, exceedingly careful studies [4].

The succession of phases suggested by the isotherms for pentadecanoic acid is typical of those found in many other systems. A generalized phase diagram can be drawn for fatty acids, for example, in which the characteristic features shift upward in temperature by 8-10K for an increase in the chain length of the tail by one $-CH_2$ unit.

The character of the LE-G and LE-LC transitions can be determined unequivocally by visual observation of the monolayer by fluorescence microscopy. In this technique [5,6] a small amount (0.1 to 2 mol %) of a fluorescent probe is added to the monolayer; typical probes are amphiphiles to which a dye such as NBD (nitrobenz-2-oxa-1,3-diazole) has been chemically grafted. The contrast between phases seen in the microscope can arise from differences in the density of the phases, differential solubility of the probe, and a quenching of the fluorescence related to changes in the orientation of the probe with respect to the water surface.

Some examples of the microscope images are given in Fig. 3, which shows the succession of monolayer textures observed when pentadecanoic acid is compressed at 25 ^{o}C. These are photographs of a television monitor on which the output of the high-sensitivity camera used to record the images is displayed. The distinctions between one-phase and two-phase regions are clearly seen. Quantitative information can be obtained by computer image analysis. For example, the fraction of the total area representing one of the phases can be measured at different average densities, allowing the densities of the coexisting phases to be determined by application of the "lever rule" [8].

The way in which such structures develop and evolve is shown in Fig. 4, which has been taken from the experiments by Stine, et al.[9] on pentadecanoic acid. One starts with the system entirely in the LE phase and then expands rapidly along a roughly isothermal path into the two-phase region; alternatively, a temperature quench can be performed at constant average density.

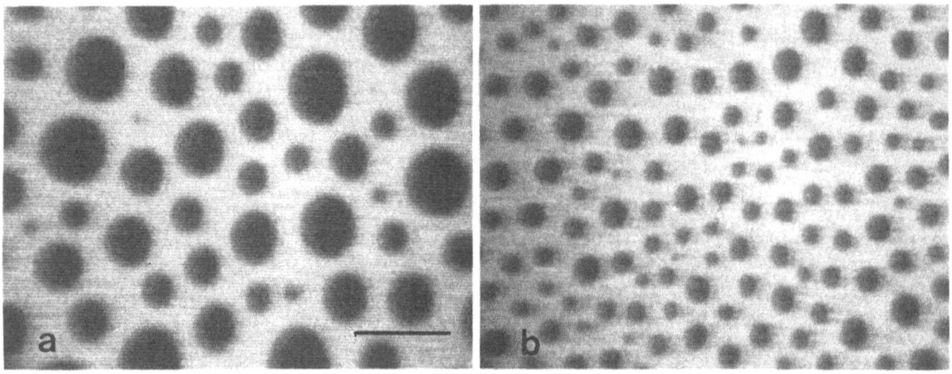

Fig. 3. Typical fluorescence microscope images observed during the compression of pentadecanoic acid. a) G-LE coexistence. The dark regions are the gaseous phase (in which the probe fluorescence is quenched) and the bright regions are the LE phase. b) LE-LC coexistence. The LC regions appear dark because the probe is not appreciably soluble in them. The bar represents 100μm.

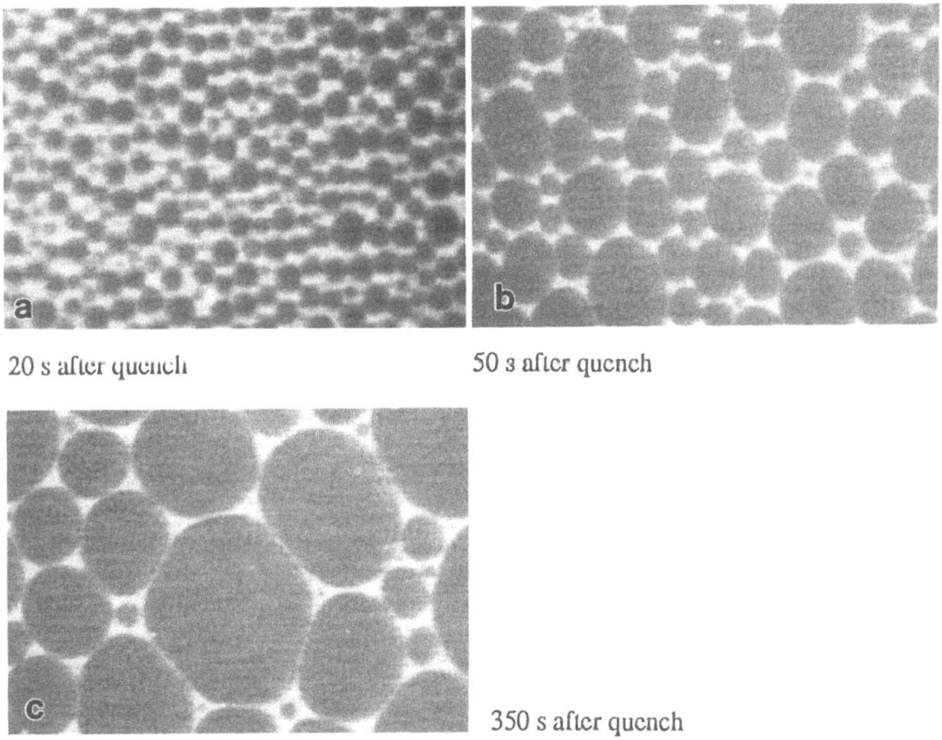

Fig. 4. Foam structures in the LE-G region of pentadecanoic acid. The microscope images show the evolution of the foam from a random array of droplets into a cellular structure in which the cells grow larger with time.

Immediately after the expansion, small dark regions appear in the initially all-white LE phase; these are bubbles of gas. The bubbles retain their circular shape until they become large enough to interact. The boundaries between them flatten and Plateau Borders [1], the triangular regions at the intersections of three bubbles, become evident. At the same time, the liquid layer between bubbles thins to about 1μm, eventually leaving a random network structure with trigonal vertices that continues to evolve.

3. Phase Diagram of Pentadecanoic Acid

From a combination of isotherm and fluorescence measurements it is possible to identify the phases and locate the phase boundaries in pentadecanoic acid [8] with some certainty. However, the precision of these measurements is still crude in comparison with those that are routinely carried out on three-dimensional systems. The molecular areas of the coexisting phases at several temperatures are given in Table I.

Table I. Coexisting Phases in Pentadecanoic Acid

	G - LE Transition		LE-LC Transition	
T/°C	$A_G^{a,b}$	A_{LE}^c	A_{LE}^c	A_{LC}^c
30	1200	44	28	21
20	1500	42	37	23

a. Area in Å^2/molecule; b. Reference 4; c. Reference 8.

The terminations of the two-phase regions have not been determined. On the basis of isotherm measurements, it had been reported [9,10] that G-LE coexistence ends at a critical point near 26 °C and that the critical point is classical, *i.e.* that the exponents have mean-field values. More recent isotherm studies [4] seem to indicate that the critical point must lie at a temperature above 40 °C, a result that is supported by fluorescence measurements [8].

The microscopic nature of the LC phase is unknown. It is possible that despite its name it is a solid phase with a symmetry different than that of the LE phase. If so, the LE-LC coexistence region cannot end in an ordinary critical point. It has been proposed that a second-order transition intercepts the coexistence curve or, as a special case of this behavior, that the second-order line meets the LE-LC coexistence curve at its maximum so that there is a tricritical point. (This is an intriguing possibility. The marginal dimensionality for tricritical points is three, hence a two-dimensional tricritical point would have nonclassical exponents.) Neither the isotherm nor fluorescence studies have yet been able to distinguish these possibilities.

4. Dynamics of Foams in Two Dimensions

A surprising result of the first investigation of monolayers of fatty acids by fluorescence microscopy [7] was the existence of foam-like structures in the region of coexistence between a gas and a condensed phase. Such structures have been observed in a variety of other systems and the phenomenon appears to be quite general.

Although the foam is long-lived it is not stable. After about four hours it begins to break down into islands of (white) droplets of the minority LE phase surrounded by the (black) gaseous phase. One can ask why the foam arises at all. It is known that *pure* three-dimensional fluids do not foam, which suggests that the monolayer foam is an artefact of the presence of the fluorescent probe. It has been observed, however, that the foam structure arises over a wide range of probe concentrations and it is seen as well [10] in monolayers consisting entirely of labeled molecules.

A key requirement for stabilizing a foam structure is a resistance to thinning of the cell edges. The amphiphiles that comprise the monolayer are dipolar and in the LE phase they are oriented out of the plane of the surface. The interactions between such pinned dipoles are repulsive and minimized by aligning the dipoles in long chains. A counterbalancing force is the line tension of the two-dimensional fluid, which acts to minimize the perimeter. McConnell and coworkers [11] have shown that a key parameter controlling the shape of a monolayer phase is λ/μ^2, the ratio of the line tension to the square of the molecular dipole moment, and that elongated domains can be more stable than more compact structures.

A random cellular structure can be characterized by the distributions for the number of sides of the cells, their sizes, local correlations of these properties, and by average and local growth rates. Several processes have been described [12] by which such structures grow: a T1 process, in which a cell edge shrinks to zero and is replaced by one in such a way that the vertices are rearranged; a T2 process, in which a cell vanishes when its edges collapse to a vertex; and inverse mitosis in which a wall separating two cells breaks and shrinks into a side. In the pentadecanoic acid foams the ratio of the T1 to T2 processes is 3:1 and inverse mitosis is rarely seen (it was more evident in stearic acid foams [7]).

The growth rate of the average cell diameter L can be examined by plotting $L - L_0$ ($=\Delta L$) as a function of $t - t_0$ ($=\Delta t$), where t_0 is the time at which the foam structure was first observed and L_0 is the average diameter at that time. A typical log-log plot of ΔL against Δt is shown in Fig. 5, where it is seen that a power law with an exponent of 0.36 0.01 gives a good representation of the data. The values of growth exponents that we have observed for pentadecanoic acid foams fall between 0.3 and 0.4,well below the value of 0.49 0.03 found in Monte Carlo simulations of foams [13] but closer to the value 0.38 obtained from simulations of grain growth [14]; they are compatible as well with the experiments by Glazier, et al. [15] on soap bubbles trapped between narrowly spaced glass plates.

A correlation between the number of sides of a cell and the average number of sides of its neighbors was determined empirically by Aboav [16] from an analysis of soap foams and was later derived by Lambert and Weaire [17] for a random arrangement of cells. It has the form

$$mn = (6 - a)n + (6a + \mu_2) \quad ,$$

where m is the average number of sides of the neighbors of a cell of n sides, μ_2 is the second moment of the distribution of n, and a is a constant that depends on the processes that give rise to the network. The time average of mn for the monolayer foam is plotted against n in Fig. 6. The line shown is a linear fit to the data for $n = 3$ to 10, which gives

$$mn = (4.74 \pm 0.04)n + (8.6 \pm 0.3) \quad . \qquad (1)$$

The values of a and μ_2 derived from this expression are $a = 1.26 \pm 0.04$ and $\mu_2 = 1.06 \pm 0.54$.

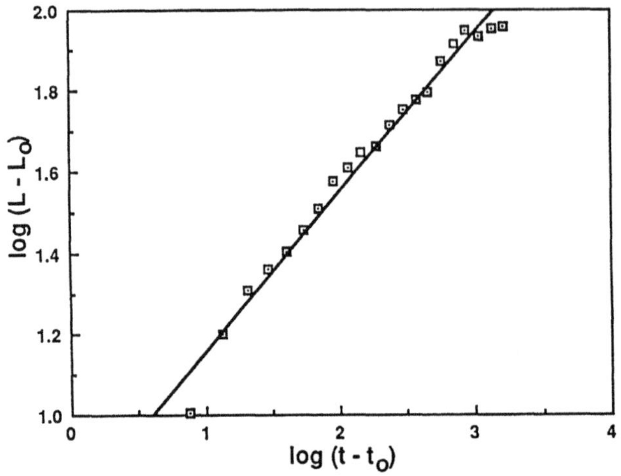

Fig. 5. Typical plot of Log ($\Delta L/\mu m$) as a function of log ($\Delta t/s$). For this run $L_0 = 24.5\ \mu m$ and $t_0 = 12.0$ s. A least-squares fit gives a slope of 0.36 $\pm$ 0.01. Other choices of L_0 and t_0 consistent with the experiment change the slope by no more than $\pm$0.02.

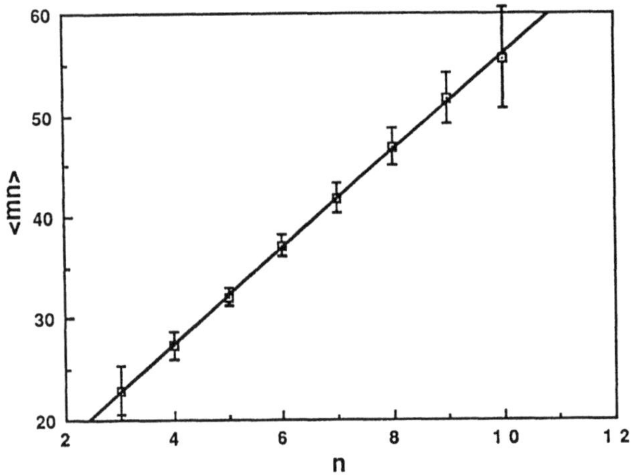

Fig. 6. Test of Aboav's Law. The ordinate is the time average of *mn*, where *m* is the average number of sides of the neighbors of a cell with *n* sides. The line has a slope of 4.74 $\pm$ 0.04 and an intercept of 8.6 $\pm$ 0.3.

An analysis of the distribution of *n* shows that μ_2 is essentially independent of time and has the value $\mu_2 = 1.38 \pm 0.24$, which is consistent with the second moment obtained from (1). The constancy of the second moment is consistent with a model of growth in which the distribution of cell sizes remains constant while the linear scale of the foam grows with time. In contrast, Weaire and Rivier [12] found that μ_2 grew linearly with time for a quasi-two-dimensional soap foam. Glazier, et al. [15] report that in their soap-foam experiments the second moment grew at early times but then leveled off to a constant value $\mu_2 = 1.4 \pm 0.1$, a result quite similar to ours. Stationary distributions with comparable second moments are also found in simulations of network structures [18].

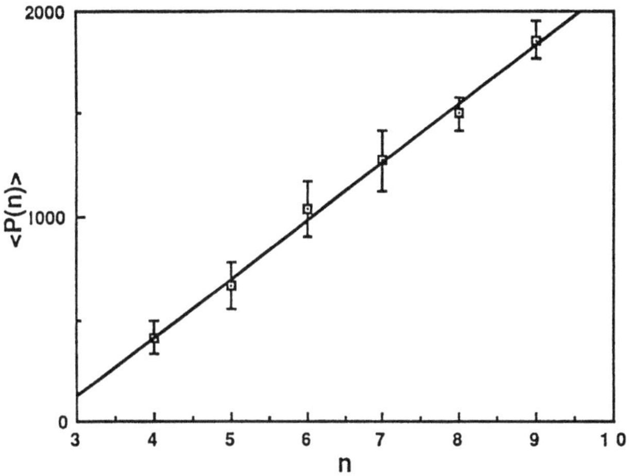

Fig. 7. Time average of the perimeter of cells as a function of the number of sides.

If the foam is a two-dimensional random space-filling structure, its character can be determined from the condition of maximum entropy [19], in which case $A(n)$, the mean area of an n-sided cell, depends linearly on n, a relation known as Lewis' law. If in addition, one imposes the constraint of conservation of the average surface energy per cell, it is found that it is the average perimeter $P(n)$ rather than the average area that is linear in n. A test of this perimeter law is shown in Fig. 7, where it is seen to be obeyed.

5. Patterns and Instabilities in Monolayers

Much of the study of pattern formation and instabilities in monolayers has focused on the LE-LC transition in phospholipids and has been carried out by McConnell and Möhwald and their collaborators [11,20]. Many of the phenomena are not unique to phospholipids and have been seen in fatty acids [21]; there are as well some patterns that have been seen in fatty acid and ester systems that have not yet been reported for phospholipids.

A key question that has not yet been answered is to what extent patterns in monolayers are equilibrium phenomena. Fractal-like structures in phospholipids [20] and fatty acids [21] appear on rapid compression, temperature quenches, or shortly after deposition of the monolayer [22], but these eventually heal into more uniform domains and are clearly non-equilibrium structures. On the other hand, Andelman, et al. [23], who have used a Ginzburg-Landau model to investigate phase transitions in monolayers of polar molecules, predict the existence of stable striped and hexagonal phases. Striated patterns and almost regular hexagonal patterns *are* seen at LE-LC transitions, but it has not been demonstrated that these patterns represent stable phases or slowly evolving patterns.

As noted before, dipolar interactions tend to stabilize linear structures in monolayers. This is demonstrated by the appearance at the G-LE transition of a buckling instability in the foam structure. The instability arises when a monolayer in the LE-G region at an average density greater than the critical density is heated. If the monolayer exhibits a foam structure it abruptly buckles as shown in Fig. 8.

The instability arises, we believe [24], because the thickness of the foam cell edges is limited by the balance between dipolar repulsions and line tension. As the temperature

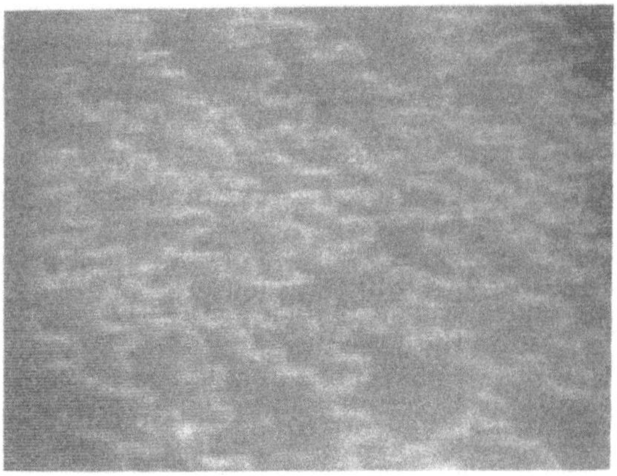

Fig. 8. Buckling instability in the foam structure. The amphiphile is ethyl heptadecanoate.

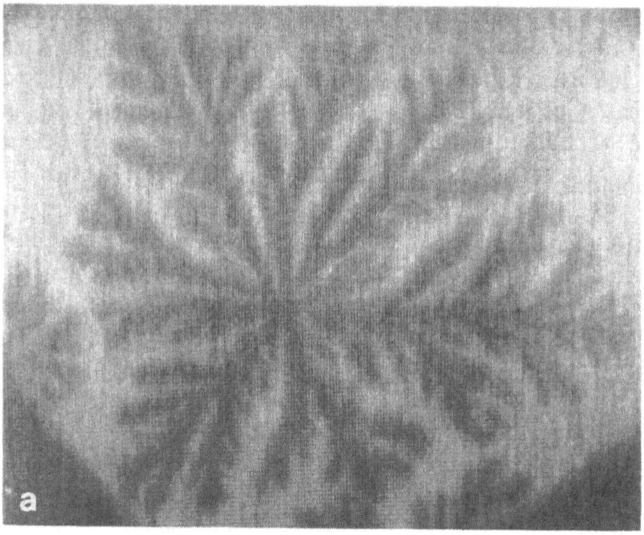

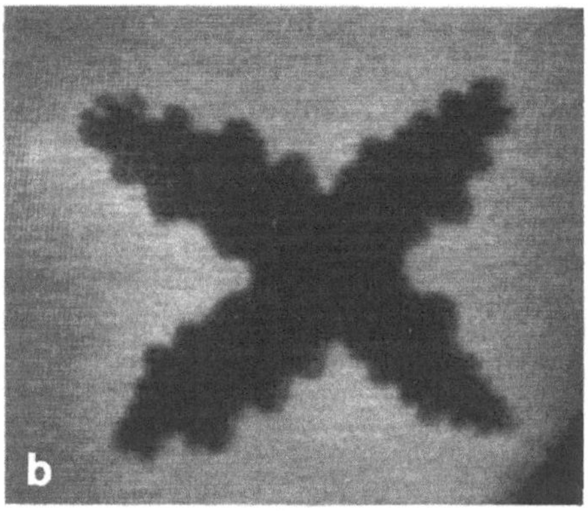

Fig. 9. Two types of dendritic growth in monolayers of ethyl octadecanoate at the LE-LC transition. a) Growth by tip splitting. b) Growth with a stable tip.

increases, the lever rule requires that the fraction of LE phase grow. Because the thickness of the edges is held constant, their length must increase, but the interconnectivity of the foam pattern will not allow this. As a result, the edges buckle, in much the same way that a rod subjected to compression can buckle [25].

By changing the nature of the monolayer it is possible to change the character of an instability. For example, we have been unable to observe dendritic growth with a stable tip in monolayers of fatty acids. As shown in Fig. 9, this type of growth can be found in esters, however, suggesting that the addition of even a small side chain provides the necessary aniosotropy in the line tension. The LC phase of esters also undergoes a shape transition as the temperature is lowered, Fig. 10. In accord with the theoretical expectations for the shapes of two-dimensional crystals [26], the domains are not facetted. The shape transition and the change in the mechanism of dendritic growth occur at roughly the same temperature and therefore probably stem from a change in the structure of the LC phase, perhaps a transition from a rotator to a non-rotator phase.

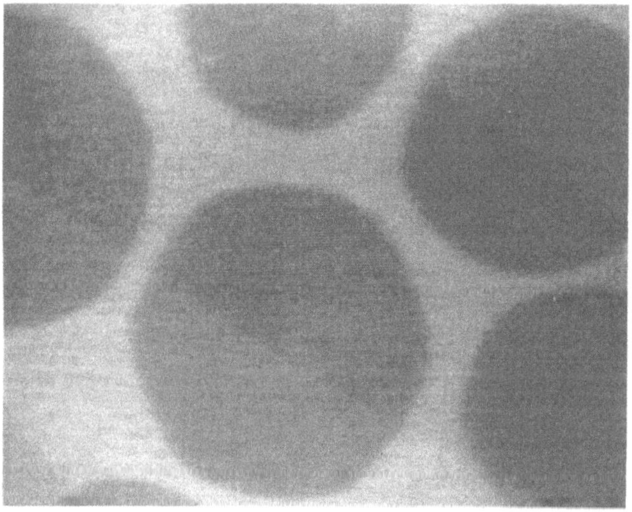

Fig. 10. Shape transition in methyl palmitate. The domains are originally circular but become hexagonal when the temperature is lowered.

Acknowledgements

We thank Steve Rauseo and Joeseph Wise for their assistance with the foam experiments. The research was sponsored by the U. S. National Science Foundation.

References

1. A. W. Adamson, *Physical Chemistry of Surfaces* (Wiley, New York, 1982), 4th ed.
2. G. L. Gaines, Jr., *Insoluble Monolayers at Liquid-Gas Interfaces* (Interscience, New York, 1966).

3. C. M. Knobler, Adv. Chem. Phys. **77**, in press.
4. N. R. Pallas and B. A. Pethica, Langmuir **1**, 509 (1985).
5. V. von Tscharner and H. M. McConnell, Biophys. J. **36**, 409 (1981).
6. M. Lösche and H. Möhwald, Rev. Sci. Instr. **55**, 1968 (1984).
7. B. Moore, C. M. Knobler, D. Broseta, and F. Rondelez, J. Chem. Soc. Faraday Trans. II **82**, 1753 (1986).
8. B. G. Moore, C. M. Knobler, S. Akamatsu, and F. Rondelez, submitted to J. Phys.Chem.
9. K. J. Stine, S. A. Rauseo, B. G. Moore, J. A. Wise, and C. M. Knobler, in preparation.
10. S. Akamatsu and F. Rondelez, personal communication.
11. See, for example, H. M. McConnell and V. T. Moy, J. Phys. Chem. **92**, 4520 (1988).
12. D. Weaire and N. Rivier, Contemp. Phys. **25**, 59 (1984).
13. D. Weaire and J. P. Kermode, Phil. Mag. B **50**, 379 (1984).
14. D. J. Srolovitz, M. P. Anderson, G. S. Grest, and P. Sahni, Scripta Met. **17**, 241 (1983).
15. J. A. Glazier, S. P. Gross, and J. Stavans, Phys. Rev. A **36**, 306 (1987).
16. D. A. Aboav, Metallography **13**, 43 (1980).
17. C. J. Lambert and D. L. Weaire, Metallography **14**, 307 (1981).
18. See, for example, K. Nakashima, T. Nagai, and K. Kawasaki, J. Stat. Phys., in press.
19. N. Rivier, Phil. Mag. B **52**, 795 (1985).
20. W. M. Heckl and H. Möhwald, Ber. Bunsenges. Phys. Chem. **90**, 1159 (1986).
21. K. A. Suresh, J. Nittman, and F. Rondelez, Europhys. Lett. **6**, 437 (1988).
22. B. G. Moore, Ph.D. dissertation, UCLA, 1989.
23. D. Andelman, F. Brochard, and J.-F. Joanny, J. Chem. Phys. **86**, 3673 (1987).
24. K. Stine and C. M. Knobler, unpublished.
25. L. D. Landau and E. M. Lifshitz, *Theory of Elasticity* (Pergamon, New York, 1970), 2nd ed., p. 97.
26. C. Rottman and M. Wortis, Phys. Rep. **103**, 59 (1984).

Geometrical Quasiparticle Condensation Model of Melting in Two Dimensions

M.A. Glaser, N.A. Clark, A.J. Armstrong, and P.D. Beale

Condensed Matter Laboratory, Department of Physics, University of Colorado, Boulder, CO 80309, USA

In this paper we describe a new model of the two-dimensional (2D) melting transition, motivated by the recent observation of Glaser and Clark that the structure of a 2D simple dense liquid can be characterized as a random square-triangle tiling containing numerous low-strength tiling faults. The model is based on the assumption that melting is associated with the proliferation and condensation of geometrical quasiparticles, which in 2D are local, nontopological fluctuations from the triangular crystal lattice to the square lattice. The quasiparticles interact via strong, highly anisotropic, short-ranged interactions that produce the characteristic tiling structure of the 2D liquid. This model is derived from and illustrated by molecular dynamics simulation studies of 2D liquid structure, and quantified using a lattice representation that embodies its essential features. The model exhibits a first-order transition for sufficiently strong quasiparticle interactions, and accounts for the volume change associated with melting in a straightforward way. However, there is no phase transition as a function of temperature for the choice of quasiparticle interactions that most accurately reproduces the local geometry of the 2D liquid, suggesting that the contribution of topological defects to the thermodynamics of the melting transition cannot be neglected.

1 Introduction

The two-dimensional (2D) solid phase is characterized by long-range bond-orientational order and quasi-long-range (algebraic) translational order. As shown by Mermin and Wagner [1], lattice vibrations are responsible for the lack of true long-range translational order in 2D solids. Recent studies of the 2D melting transition have focused on the loss of translational and bond-orientational order upon moving from the low-temperature solid to the high-temperature liquid phase [2]. Because the symmetry changes associated with melting depend on the formation of unbound topological defects (dislocations and disclinations), a number of topological-defect-mediated melting models have been proposed [2].

In this paper we describe a model for 2D melting derived from molecular dynamics (MD) simulation studies of 2D simple dense liquids and solids. Our MD simulations demonstrate that an important aspect of the melting transition is the appearance of geometrical (nontopological) disorder in the liquid phase [3]. This disorder manifests itself in a significant degree of square lattice coordination in the liquid phase, with particles adopting local arrangements characteristic of plane tilings composed of squares and equilateral triangles (ST tilings). These observations form the basis for our model of 2D melting, in which the transition is mediated by nontopological (geometrical) defects, or quasiparticles, envisioned as local "square" distortions of a triangular lattice. The quasiparticles interact via short-ranged, anisotropic, attractive interactions. Sufficiently strong quasiparticle interactions produce a first-order condensation transition to a disordered phase exhibiting some of the the key structural features of the simulated liquid near freezing. The proliferation of quasiparticles upon melting accounts straightforwardly for the volume change associated with melting. Our results suggest, however, that the contribution of topological defects to the thermodynamics of melting cannot be neglected. Our model is related to one originally proposed by Collins [4], and later extended by Kawamura [5] and others [6] — their tiling model represents 2D liquid structure in terms of disclination-free random ST tilings, i.e. tilings composed of perfect squares and equilateral triangles. Here we show that, although local 2D liquid structure is usefully described in terms of ST tilings, the restriction to disclination-free tilings is too limiting.

2 Statistical Geometry of 2D Liquids

As described previously [3], we carried out MD simulations of systems of 896 and 3584 particles, interacting via a truncated Lennard-Jones (WCA) pair potential [7] at a reduced temperature of $T = 0.6$ [8] and number densities in the range $0.70 \leq \rho \leq 0.91$. At this temperature we found a first order melting transition with two-phase coexistence in the range $0.85 \lesssim \rho \lesssim 0.88$ [3]. To assess the local structure of the 2D liquid we suppressed vibrational fluctuations by averaging particle configurations over a time window of 1.0 Lennard-Jones (LJ) units, comparable to the period of single-particle vibrational fluctuations but short compared to the characteristic time for large-scale structural change.

We developed a variety of methods for characterizing the local geometrical structure of the 2D WCA system, all of which are derived from the Voronoi construction [9]. Pairs of particles whose Voronoi cells share a common edge are taken to be nearest neighbors. Once we have defined a set of nearest neighbors for each particle in this way, we can probe local solid-like order in the liquid by computing the local sixfold bond-orientational order parameter

$$\psi \equiv \psi_i = \frac{1}{n_i} \sum_{j=1}^{n_i} e^{i6\theta_{i,j}}. \tag{1}$$

The sum runs over the n_i nearest neighbors of particle i, and $\theta_{i,j}$ is the orientation of the nearest-neighbor bond between particle i and its jth neighbor relative to an arbitrary direction. We identify solid-like regions, or "triangular ordered clusters" (TOCs) by selecting particles having $|\psi| \geq 0.75$ — nearest-neighbor pairs of particles that both satisfy this criterion are assigned to the same TOC. In Fig. 1a we have plotted the Voronoi cells of particles in TOCs for a typical time-averaged configuration of the WCA liquid at coexistence ($\rho = 0.85$), with cell boundaries interior to TOCs in green, exterior boundaries in blue, and particle positions marked by blue dots. The WCA liquid near freezing clearly exhibits extensive solid-like ordering — indeed, a "percolating cluster" is evident in this configuration. In Fig. 1a, we have also shown nearest-neighbor bond angles that exceed 75° (red lines). This illustrates a striking feature of the 2D liquid — the regions surrounding TOCs contain numerous local groups of particles having a nearly square lattice arrangement. Such "squares" typically form linear chains, or "ladders."

As shown previously [3], local fluctuations to the square lattice give rise to distinct peaks in the nearest-neighbor bond-length and bond-angle distribution functions. This fact enables us to use a simple bond-dilution procedure (the "polygon construction") to identify local square lattice structure [3]. The network of nearest-neighbor bonds obtained from the Voronoi construction partitions the plane into triangular areas. Removal of nearest-neighbor bonds associated with the secondary peaks in the bond-angle or bond-length distribution functions results in a partitioning of the plane into polygonal areas, or "tiles," having three or more sides (in the examples we show here, the bond-dilution procedure involves the removal of bonds opposite nearest-neighbor bond angles larger than 75°; alternatively, we could remove bonds that are significantly longer than the most probable nearest-neighbor distance). For quantitative comparison with ST tilings, polygons having more than four sides can be "reduced" by successively replacing their shortest internal bonds until they are broken into three- and four-sided polygons. The resulting "reduced polygon construction" is shown in Fig. 1b, for the configuration of Fig. 1a. The bond-dilution procedure leads to the removal of diagonal bonds in groups of four particles with nearly square structure, yielding four-sided polygons (shown in red). Regions that consist primarily of triangles (shown in green) are associated with TOCs (compare with Fig. 1a).

The local arrangements of "squares" (four-sided polygons) and triangles in the 2D liquid strongly resemble those found in ST tilings. A typical ST tiling is shown in Fig. 1d. ST tilings form the basis of several 2D melting models [4-6], and, more recently, have been discussed as models for 2D dodecagonal quasicrystals [10-13]. ST tilings are a mosaic of the four types of vertex that satisfy the tiling rules for squares and equilateral triangles (inset, Fig. 1d). As is evident from the figure, the local constraints (tiling rules) produce extended correlations that are responsible for the lizard-skin-like texture of the tiling. We will occasionally describe ST tilings as "ideal" tilings, referring to the absence of tiling faults (disclinations) in such tilings.

To compare the local geometry of the 2D liquid with that of an ideal ST tiling, we can classify vertices in the reduced polygon construction either as one of the four types allowed under the ST tiling rules ("ST-allowed" vertices) or as tiling faults (disclinations) [14]. In Fig. 1c we show an example of such a vertex classification for the configuration of Fig. 1b, with ST-allowed vertices marked with the corresponding symbol from the inset of Fig. 1d and tiling disclinations indicated by plus or minus signs. The vast majority of the non-ST-allowed vertices are +1 (−1) disclinations in the tiling [14]: vertices surrounded by five (four) triangles and a square, i.e having a 30° excess (deficit) in angle. Of the 896 vertices in this configuration, 879 of them (98%) are either ST-allowed (63%) or unit strength disclinations (35%). Sixteen of the remaining 17 vertices are of strength 2, and the remaining vertex is a rare +3 disclination.

Clearly there is a tendency for particles in the dense 2D liquid to form local arrangements characteristic of ST tilings. This tendency, which is evident for a wide range of densities near freezing, appears to be a result of packing constraints [3]. A local ST-tiling structure is favorable because it provides relatively efficient packing and gives a finite entropy contribution due to the many possible arrangements of "tiles."

3 Geometrical Quasiparticle Condensation (GQC) Model of Melting

Our model for 2D melting can be simply described with the aid of Fig. 1b. We consider the red polygons to be "geometrical" (nontopological) quasiparticles. These quasiparticles are assumed to require some creation energy and to interact via short-ranged, anisotropic, attractive forces that produce the the local ST-tiling structure observed in the 2D liquid. For sufficiently strong attractions the model can exhibit a first-order "condensation" transition from a low-temperature phase containing few quasiparticles (solid phase) to a high-temperature phase containing a high density of quasiparticles (liquid phase), in which the quasiparticles aggregate into local ST-tiling structures. We identify this condensation transition with the melting transition. The order parameter (really a disorder parameter) for the condensation transition is the quasiparticle density, $d_q = N_q/N$, where N_q is the number of quasiparticles and N is the number of particles. We will refer to this general melting mechanism as "geometrical quasiparticle condensation", or GQC.

The quasiparticles that mediate the transition are thermally excited, nonlinear distortions of the ideal crystal lattice, which can be generated by continuous displacement of local groups of particles. The quasiparticles are nontopological defects that do not always show up in analysis of solid and liquid structure in terms of coordination number, i.e. by identifying dislocations or disclinations — they characterize local geometrical rather than topological structure. Fig. 1e shows the vertex classification for an isolated 2D quasiparticle in the WCA solid near melting ($\rho = 0.885$). An isolated quasiparticle is a local fluctuation of a group of four neighboring particles at the vertices of a 60° rhombus to form a square lattice unit cell of the same lattice parameter, and is associated with four tiling disclinations of alternating sign (a disclination quadrupole).

A 2D geometrical quasiparticle can be regarded as a geometrical excitation from the triangular lattice "vacuum state." Associated with such excitations is a creation energy E_c resulting from distortion of the lattice, which depends on the form of intermolecular potential and the elastic properties of the lattice. Once formed, the quasiparticles interact via long-range elastic interactions, which we neglect, and via short-range, anisotropic, attractive forces that favor aggregation into specific (ST tiling) structures and that favor the appearance of additional quasiparticles once some are present.

To develop this qualitative picture into a model for melting, the quasiparticle interactions must be specified. In 2D these interactions clearly must favor aggregation into ST tiling structures, since these are locally prevalent in the liquid. Ideal ST tilings exhibit a variety of characteristic structures including ladder-like rows of squares (quasiparticles) intersecting in groups of three or four [5]. The quasiparticle interactions must be chosen so that quasiparticles are stabilized by aggregation into the local structures characteristic of ST tilings.

Quasiparticle interactions that stabilize local ST tiling structures can be formulated as follows. Although the quasiparticles have no net topological strength, they can be usefully associated with the disclinations in the ideal ST tiling as discussed above. As shown in Fig. 1e, an isolated quasiparticle is a disclination quadrupole in the ST tiling, containing two +1 (five triangles plus a square) and two −1 (four triangles and a square) disclinations. When quasiparticles aggregate into local ST tiling structures, tiling disclinations on neighbors cancel to form allowed (strength 0) vertices. This suggests that quasiparticle interactions favoring a local ST tiling structure can be obtained by assigning a positive creation energy E_d to each disclination. The energy associated with tiling disclinations is minimized when quasiparticles meet with the correct geometry (Figs. 1f-h). The formation of "ladders" (Figs. 1f,h) is particularly favorable since linear chains of quasiparticles can be extended indefinitely without the creation of new disclinations.

The creation of an isolated quasiparticle involves a local expansion of the triangular lattice, giving quasiparticles something of the character of voids. This property suggests that the volume increase associated with melting is related to the proliferation of quasiparticles. For the 2D WCA system, the fractional change in specific area on melting is $\Delta a/a_s = 0.035$, where a_s is the area per particle in the solid at coexistence. The theoretical quasiparticle contribution to $\Delta a/a_s$ can be obtained from

$$(\Delta a/a_s)_q = (d_{ql} - d_{qs})\delta a_q \tag{2}$$

where d_{ql} and d_{qs} are the quasiparticle densities in the liquid and solid at coexistence, respectively, and $\delta a_q = 2/\sqrt{3} - 1$ is the fractional area expansion associated with the deformation of a 60° rhombus into a square. For the WCA system we find $d_{ql} = 0.244$ and $d_{qs} = 0.026$ [15], which gives $(\Delta a/a_s)_q = 0.034$, in good agreement with the observed value of 0.035. Thus the GQC mechanism appears to account for the volume change associated with melting in a straightforward way.

4 Lattice Version of the GQC Model

In order to quantitatively assess the properties of this quasiparticle system, we have formulated a lattice model that incorporates the essential features of the GQC model. We map the quasiparticle system onto the bonds of a two-dimensional triangular lattice in a way analogous to a similar mapping of disclination-free ST tilings [5], and model the creation of each quasiparticle by the breaking of a single bond. The energy required to break an isolated bond is $E_c + 4E_d$ — less energy is required if the bond to be broken is properly situated with respect to nearby broken bonds, so that cancellation of disclinations (formation of local ST tiling structures) occurs (as in Figs. 1f-h). The Hamiltonian for this system is

$$H = E_c \sum_{i=1}^{3N} s_i + E_d \sum_{\alpha=1}^{N} n_\alpha^2 + E_r \sum_{(i,j)} s_i s_j, \tag{3}$$

where N is the number of vertices in the lattice and $3N$ is the number of bonds. The bond spin variables s_i, which reside on a Kagomé net [16], take on the values 0 or 1 (corresponding to intact and broken bonds, respectively). The vertex disclination charge n_α is not independent of the bond spins, but is related to the bond spin variables by

$$n_\alpha = -\sum_{k=1}^{6} s_k + \sum_{l=1}^{6} s_l, \tag{4}$$

where the two sums range over the first and second shells of bonds surrounding vertex α, respectively. The third term in the Hamiltonian runs over nearest-neighbor pairs of bonds. In the work described here we have chosen $E_r = +\infty$ to prevent the removal

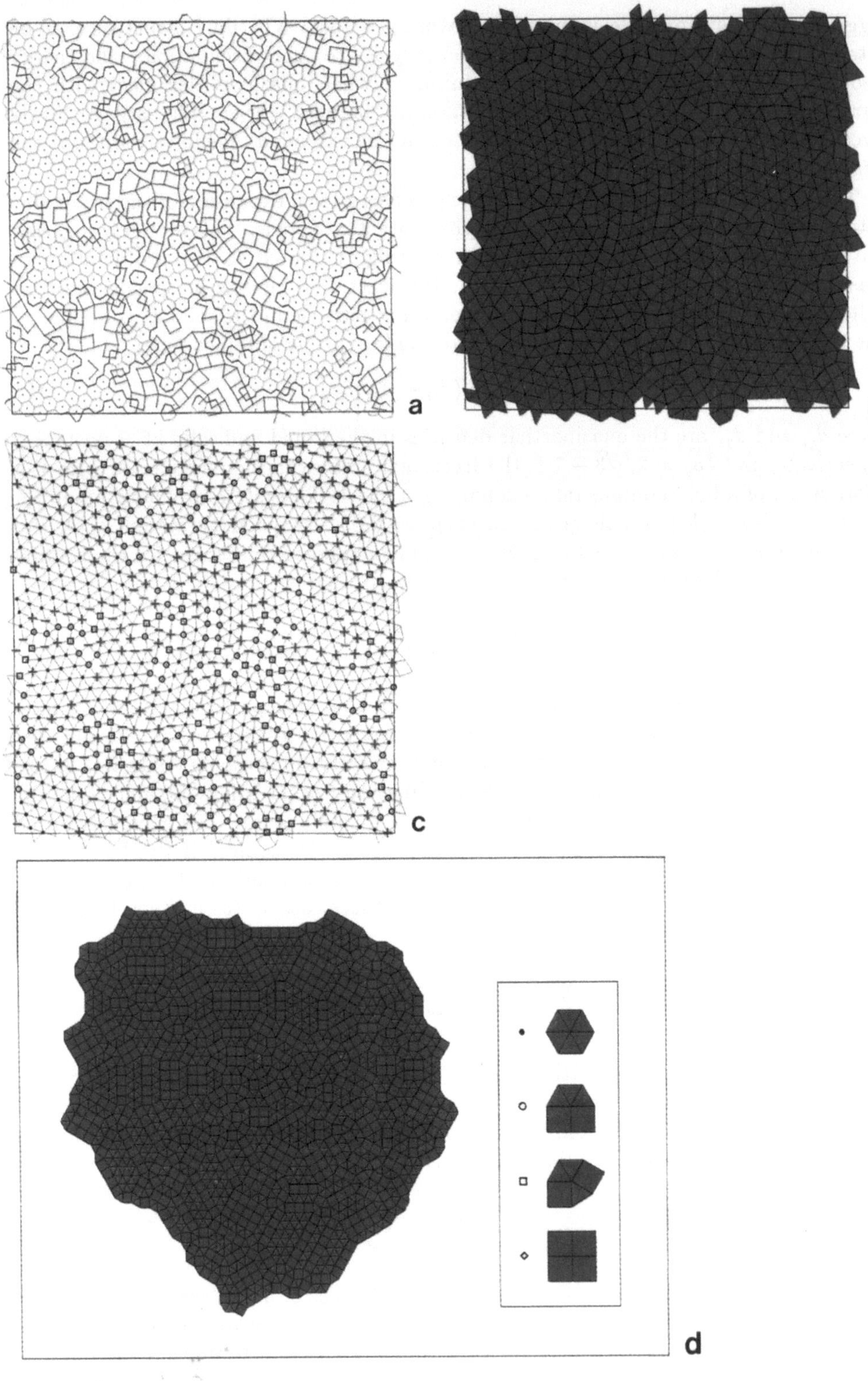
a
b
c
d

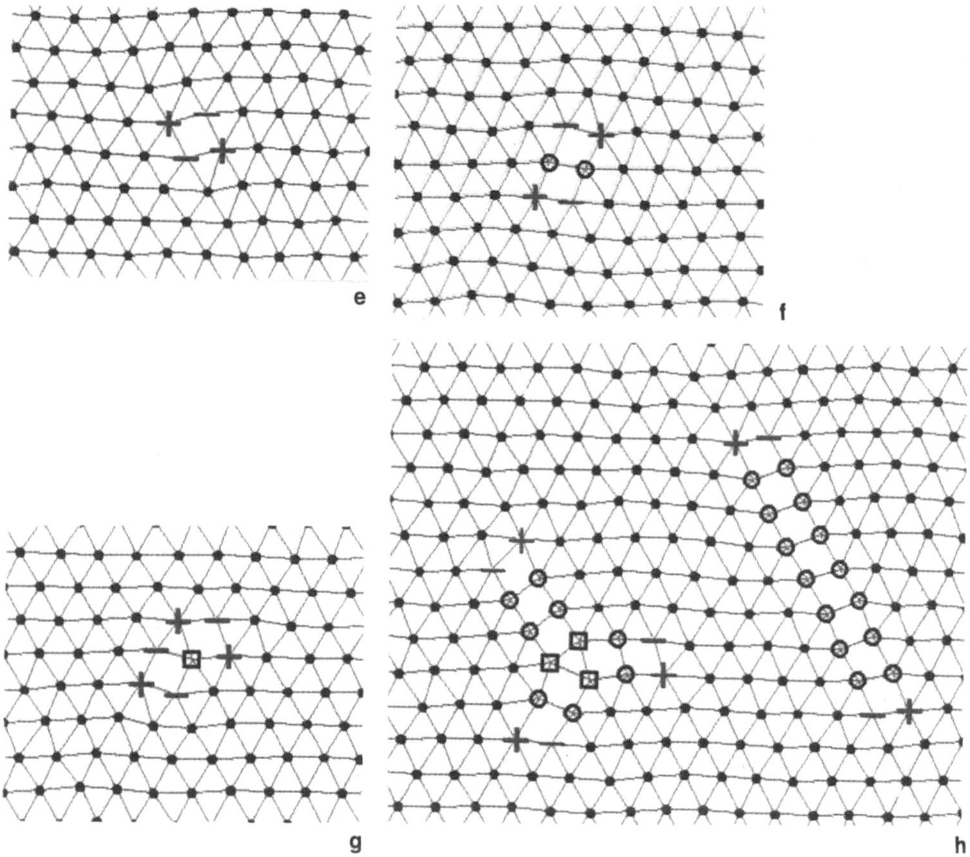

Figure 1. (a) Typical time-averaged WCA liquid configuration at $\rho = 0.85$. The Voronoi cells for particles in triangular ordered clusters (TOCs) have been drawn, with cell boundaries interior to TOCs in green and exterior boundaries in blue. Particle positions are marked by blue dots. Nearest-neighbor bond angles exceeding 75° are also shown, in red, indicating square-lattice-like coordination in non-triangular regions. (b) Reduced polygon construction for the configuration shown in (a). The internal (diagonal) nearest-neighbor bond within each group of particles having a nearly square arrangement has been deleted to form a four sided polygon. (c) Vertex classification for the configuration of (a), with ST-allowed vertices identified by the corresponding symbol from the inset of (d), and tiling disclinations marked by plus of minus signs, indicating their sign, with disclinations of strength 1 in red and those of strength ≥ 2 in cyan. (d) A representative random, ideal (disclination-free) square-triangle (ST) tiling. Inset: the four types of vertices allowed by the ST tiling rules. (e) Vertex classification for an isolated quasiparticle in the WCA solid near melting ($\rho = 0.885$). (f-h) Vertex classification for several examples of energetically favorable quasiparticle aggregates in the WCA solid near melting ($\rho = 0.885$).

of any pair of neighboring bonds (this corresponds to only allowing three- and four-sided polygons). Although the Hamiltonian has a simple form when written in terms of vertex charges, the effective bond spin interactions are extended, with the energy of a given bond depending on the state of 28 neighboring bonds. This lattice model is described by two dimensionless parameters, namely $t = T/E_c$ and $r = E_d/E_c$, where T is the temperature in energy units.

We have obtained the basic statistical mechanical properties of the quasiparticle lattice model via Metropolis Monte Carlo (MC) simulations, carried out on lattices of various sizes. Figs. 2a-i show a variety of thermodynamic properties of the lattice model for various values of t and r, obtained from MC simulations of a 3584-vertex (10752-spin) system with periodic boundary conditions. At each value of t and r, the system was thermalized for 10^4 sweeps (MC steps per bond) and thermodynamic averages were computed for the next 10^4 sweeps. The curves shown in Fig. 2 represent sequences of runs for increasing temperatures at a given r-value, with a thermalized configuration from each temperature used to start the run at the next highest temperature. The various curves are labelled by their r-values. Fig. 2a shows the order parameter (d_q), Fig. 2b the potential energy per vertex, Figs. 2c-e the fractions of charge 0, charge +1 and charge −1 disclinations, and Figs. 2f-i the fractions of the four types of ST-allowed vertex shown in the inset of Fig. 1d. Discontinuities are observed in all quantities for large r, indicative of a first-order transition. Cooling runs were also performed, and hysteresis was observed for large values of r, also suggestive of a first-order transition. If r is very large, only vertices of strength 0 (ST-allowed vertices) will appear and in

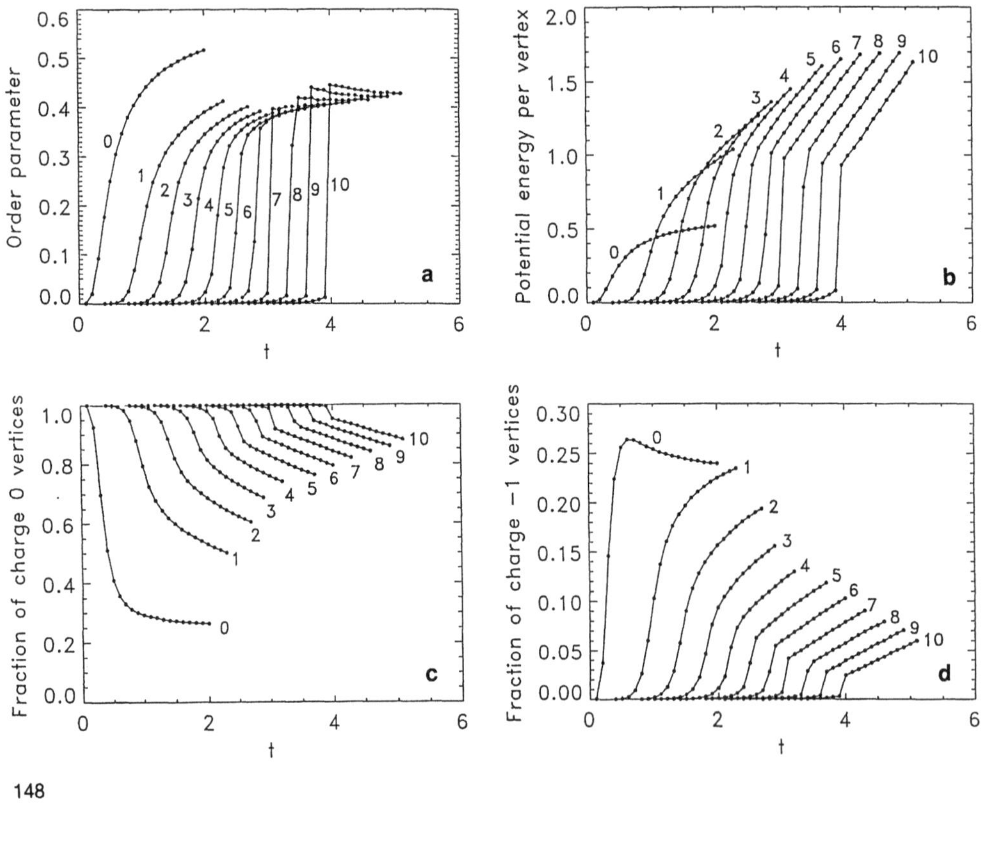

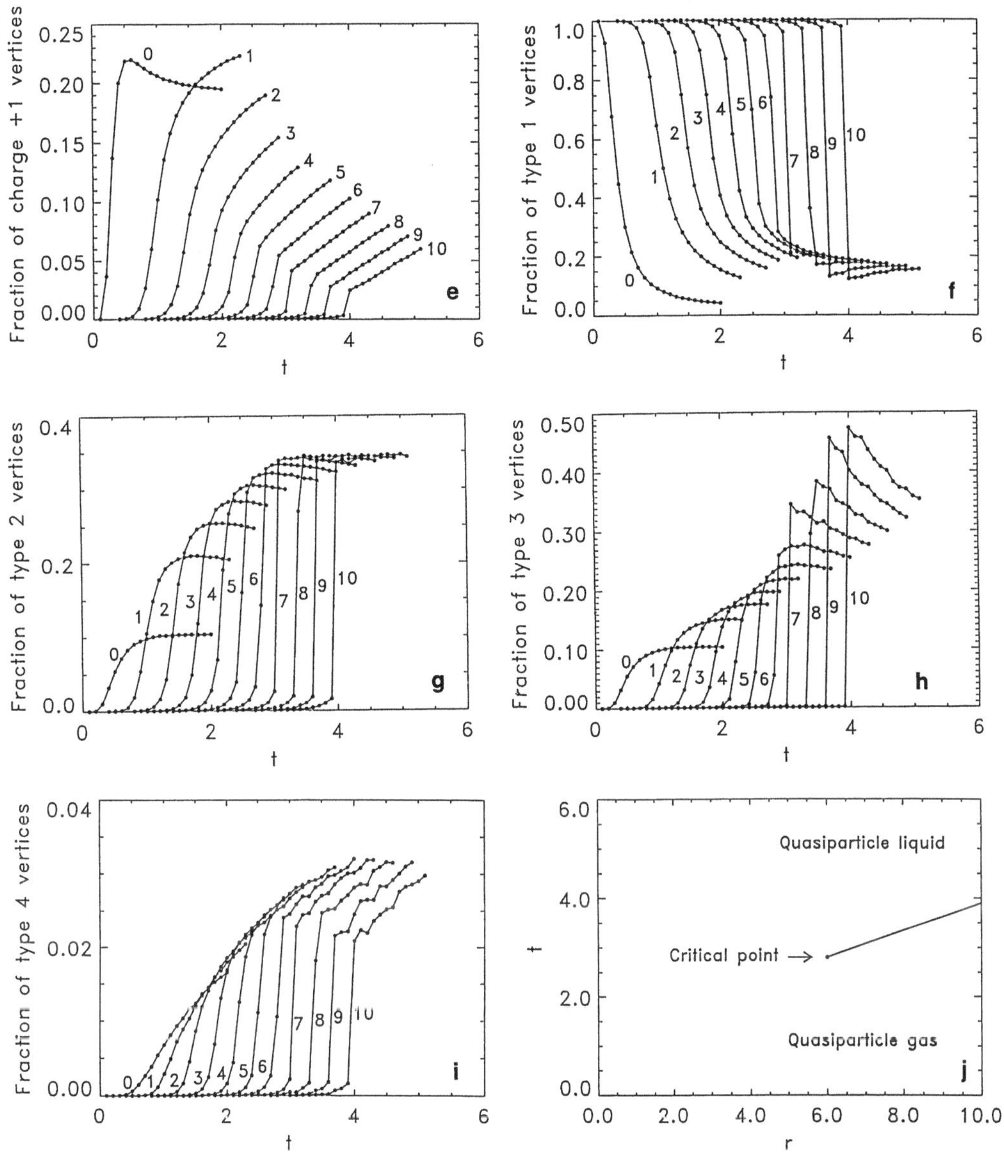

Figure 2. (a-i) A variety of properties of the the GQC lattice model obtained from Monte Carlo simulations of a 3584-vertex system. Each curve represents a series of simulations for increasing t at a given value of r, and is labelled by its corresponding r-value. (a) Order parameter (d_q). (b) Potential energy per vertex. (c) Fraction of charge 0 (ST-allowed) vertices. (d) Fraction of charge −1 disclinations. (e) Fraction of charge +1 disclinations. (f) Fraction of type 1 vertices (first type of ST-allowed vertex shown in the inset of Fig. 1d). (g) Fraction of type 2 vertices (second type of ST-allowed vertex shown in the inset of Fig. 1d). (h) Fraction of type 3 vertices (third type of ST-allowed vertex shown in the inset of Fig. 1d). (i) Fraction of type 4 vertices (fourth type of ST-allowed vertex shown in the inset of Fig. 1d). (j) Phase diagram of the GQC lattice model. The solid line is a first-order transition line.

this limit we obtain Kawamura's first-order transition for the ideal ST tiling [5]. For $r \lesssim 6$ there is no evidence of a phase transition, indicating the presence of a critical point near $r = 6$. Fig. 2j shows the (somewhat schematic) phase diagram of the GQC model.

From Figs. 2d,e and Figs. 2g,h we see that the model exhibits asymmetry between the fractions of charge +1 and charge −1 vertices, and between the fractions of type 2 and type 3 vertices, features that are also observed in the 2D WCA liquid. This gives us confidence that our model contains the essential ingredients needed to describe the local geometry of the 2D liquid.

We have found [15] that the vertex statistics of the GQC model (e.g. fraction of ST-allowed vertices, relative proportions of various types of ST-allowed vertices, etc.)

Figure 3. (a) Typical configuration of the GQC lattice model at $r = 0.8$, $t = 0.7$. Deleted bonds are indicated by red rhombi. (b) Typical configuration of the GQC model at $r = 0.8$, $t = 1.1$. (c,d) Relaxed configurations corresponding to the configurations of (a,b).

at $r = 0.8$ match those of the WCA system rather closely. Figs. 3a,b show typical MC configurations at $r = 0.8$, for $t = 0.7$ and $t = 1.1$, respectively. The interactions clearly favor the formation of the characteristic local ST structures, with rows of quasiparticles intersecting in threes and fours. A more direct structural comparison with the WCA system can be made by deforming the quasiparticles into their typical square geometry. To do this the vertices were connected by Hooke's law springs, with the springs associated with deleted bonds having an equilibrium length of $\sqrt{2}$ and those associated with intact bonds having an equilibrium length of 1, in units of the triangular lattice constant. The potential energy of this elastic network was then minimized using a conjugate-gradient procedure. The resulting structures, Figs. 3c,d, generated from the lattice configurations of Figs. 3a,b, have a local geometry that closely resembles that of the WCA solid near melting and the WCA liquid near freezing, respectively (compare Fig. 3d with Fig. 1b).

The alert reader will note that, according to the phase diagram shown in Fig. 2j, there is no phase transition as a function of t for $r = 0.8$, the r-value for which the vertex statistics of the GQC lattice model most closely match those of the WCA system. The likely implication of this fact is that topological defects (which are excluded from our lattice model by construction) play an important role in determining the thermodynamics of the melting transition. A more realistic model would allow for the creation of isolated topological defects, and would include interactions between geometrical defects and topological defects.

5 Conclusions

Generally, GQC is related to the topological-defect-mediated melting mechanisms of recent vintage [2]. However, it is distinct in several basic ways. The defects responsible for the transition in the GQC model (quasiparticles) are non-topological and interact via a highly anisotropic short-range potential related to local packing or geometric constraints rather than via far-field elastic distortions of the "vacuum." Whereas melting models based on topological defects focus on changes in the long-range order of a system to define its order parameter, and attempt to relate the thermodynamic properties of a system to its long-wavelength fluctuations, in the GQC model the emphasis is on short-wavelength fluctuations, and the order parameter (the quasiparticle density) characterizes the local geometry (rather than the topology) of 2D systems. It is clear that topological defects must be included in a proper description of the 2D melting transition. However, we feel that the contribution of short-wavelength, geometrical fluctuations to the thermodynamics of melting must also be considered.

Although at present we have carried out detailed studies only for 2D systems, the similarity of the melting phenomenology for two- and three-dimensional systems having similar interactions suggests that the GQC model may be relevant to three-dimensional systems as well. The structure of the quasiparticle in a given system will depend on the structure of the constituent particles and the spatial dimensionality. Thus we propose that, for example, three-dimensional simple liquids can be viewed as a mosaic of crystal-like and quasiparticle regions that locally fit together in the three dimensional equivalent of a tiling.

We would like to thank Rainer Malzbender and Mike Widom for helpful suggestions. This work was supported by NSF Low Temperature Physics Grant DMR 88-07443.

References

[1] N. D. Mermin and H. Wagner, Phys. Rev. Lett. **17**, 1133 (1966).

[2] Work on 2D melting has been recently reviewed by K. J. Strandburg, Rev. Mod. Phys. **60**, 161 (1988) and by H. Kleinert, *Gauge Fields in Condensed Matter* (World Scientific, 1989).

[3] M. A. Glaser and N. A. Clark, to appear in Phys. Rev. A, Rapid Comm.

[4] R. Collins, Proc. Phys. Soc. **83**, 553 (1964).

[5] H. Kawamura, Prog. Theor. Phys. **70**, 352 (1983).

[6] Y. M. Yi and Z. C. Guo, J. Phys.: Condens. Matter **1**, 1731 (1989).

[7] J. D. Weeks, D. Chandler, and H. C. Andersen, J. Chem. Phys. **54**, 5237 (1971).

[8] Unless otherwise noted, all quantities are expressed in LJ units [7], in which the LJ ϵ is the unit of energy, the LJ σ is the unit of length, and the particle mass m is the unit of mass.

[9] G. F. Voronoi, J. Reine. Agnew. Math. **134**, 198 (1908).

[10] H. Chen, D. X. Li, and K. H. Kuo, Phys. Rev. Lett. **60**, 1645 (1988).

[11] P. W. Leung, C. L. Henley, and G. V. Chester, Phys. Rev. B **39**, 446 (1989).

[12] Q. B. Yang and W. D. Wei, Phys. Rev. Lett. **58**, 1020 (1987).

[13] K. H. Kuo, Y. C. Feng, and H. Chen, Phys. Rev. Lett. **61**, 1740 (1988).

[14] We first compute the sum Σ of "nominal" bond angles around a vertex (a triangle contributes 60° and a "square" 90°). The strength of a disclination is then given by $n = (\Sigma - 360°)/30°$. ST-allowed vertices are those having $n = 0$.

[15] M. A. Glaser and N. A. Clark, to be published.

[16] F. C. Frank and J. S. Kasper, Acta Cryst. **12**, 483 (1959).

Mixed Polymer Films at the Air–Water Interface

M. Kawaguchi, R. Nishida, and K. Nagata

Department of Industrial Chemistry, Faculty of Engineering,
Mie University, 1515 Kamihama-cho, Tsu, Mie 514, Japan

We have carried out measurements, using the Wilhelmy plate method and ellipsometry, for the elucidation of the interfacial properties, such as compatibility and interfacial conformation of the binary mixtures of poly(ethylene oxide)(PEO) and poly(methyl methacrylate) (PMMA) spread at the air-water interface.

1. Surface pressure measurements

In general, the compatibility of the polymer mixed films is determined from the plots of both the mean areas at a constant surface pressure and the collapse surface pressure versus molar fraction of one component in the binary mixture/1/.

We notice some features in the surface pressure-area isotherms of the mixtures. The isotherms at higher PEO contents have plateau regions around a surface pressure of 10 mN/m, which corresponds to the collapse surface pressure of PEO film. The width of the plateau regions increases with increasing mole fraction of PEO and the surface pressure in the plateau regions increases with an increase in PEO content. This plateau surface pressure could be regarded as a kind of collapse pressure. There is the other collapse surface pressure, i.e., 15-17 mN/m at the small area. This high collapse surface pressure increases with increasing PEO content.

Plots of the mean areas at a given constant surface pressure against the molar fraction of PMMA deviate below an additivity line. The additivity line is expressed by equation 1 using molar ratios (X_1, X_2) and areas per unit mass (A_1, A_2) of the two components 1 and 2 if they are ideally miscible or completely immiscible.

$$A = X_1A_1 + X_2A_2 \tag{1}$$

where A is the mean area per unit mass at a given surface pressure. The negative deviation means that the intermolecular interaction between PEO and PMMA is attractive and the mixture of PEO and PMMA at the air/water interface is non-ideally miscible and stable, and moreover, both polymers more easily interpenetrate each other. The compatibility of both polymers is also guaranteed by the composition dependence of the plateau surface pressure as well as the higher collapse surface pressure. This compatibility of PEO and PMMA in the two dimensional state well correlates with that of the blends of PEO and PMMA in the bulk state/2/.

2. Ellipsometry

Ellipsometry gives the two quantities of the phase difference, Δ, and the azimuth, Ψ, of the amplitude ratio for light polarized parallel and normal to the plane of incident. Kawaguchi et al./3,4/ and Sauer et al./5/ found that the changes in Δ, $\delta\Delta$, between the pure water and the water surface covered by polymers was much more sensitive to the presence of the polymers at the air/water interface than those in Ψ, $\delta\Psi$, for several polymers. From calculation of the thicknesses of PEO and PMMA layers spread at the air/water interface from the ellipsometric quantities, it was concluded that not only water soluble PEO but also water insoluble PMMA were well hydrated.

The $\delta\Delta$ values for the mixtures of PEO and PMMA are also much more larger than $\delta\Psi$. The $\delta\Delta$ values as a function of the surface concentration of PMMA in the mixtures have two steps. First, there is a step around the surface PMMA concentration = 0.6-1.0 mg/m^2, where the low collapse surface pressure is observed: secondly, the $\delta\Delta$ values have a plateau value above the surface PMMA concentration of 17 mg/m^2 at which the surface pressure is in the high collapse surface pressure. Thus, there are some correlations between the $\delta\Delta$ value and the surface pressure.

For the mixtures with higher PMMA contents the observed $\delta\Delta$ is in agreement with the value calculated from the simple addition of the $\delta\Delta$ for the individual polymers at a given surface concentration. In contrast, for the mixtures with lower PMMA contents the experimental values of $\delta\Delta$ are not consistent with the additive values since the refractive index of water phase is higher than that of pure water due to desorption of PEO from the air/water interface.

Thicknesses could be calculated from only the value of $\delta\Delta$ using the method of Sauer et al./5/ because of the very small value of $\delta\Psi$. In the model of Sauer et al. the refractive index of the polymer layer is assumed to be dependent on the surface coverage of the water surface with the polymer and to be equal to that in the bulk state at the full surface coverage. Since both polymers are comparable, the refractive index of the mixed polymers could be calculated from the Lorentz-Lorenz equation. Thicknesses so calculated increase with increasing PEO content and they are larger than the thickness of PMMA.

3. References

1 G. Gabrielli, M. Puggelli and P. Baglioni, J. Colloid Interface Sci. **86**, 485 (1982).

2 E. Martuscelli, M. Pracella and P. Y. Wang, Polymer **25**, 1097 (1984).

3 M. Kawaguchi, M. Tohyama, Y. Mutoh and A. Takahashi, Langmuir **4**, 411 (1988).

4 M. Kawaguchi, M Tohyama and A. Takahashi, Langmuir **4**, 411 (1988).

5 B. B. Sauer, H. Yu, M. Yazdanian, G. Zografi and M. W. Kim, Macromolecules **22**, 2332 (1989).

Part VI

Coalescence and Domain Growth

Simple Models for Coalescence of Fluid Droplets

P. Meakin

Central Research and Development Department, E.I. du Pont de Nemours and Company, Wilmington, DE 19880, USA

Abstract. A variety of models involving the coalescence of D dimensional hyper-spherical droplets in or on d dimensional substrates have been explored using computer simulations. In all cases, the simulation results can be understood in terms of simple scaling and/or statistical models. Consequently, exact but nonrigorous values for the exponents that describe the scaling behavior have been obtained as a function of D and d. These exponents are found to be universal (insensitive to model details) so that the results described here may be relevant to real systems that are not described in full detail by the models.

1. Introduction

Processes in which the coalescence of fluid droplets play an important role are common in many areas of science and technology. Familiar examples include the condensation of water droplets on cold surfaces[1-3], rain, fog and clouds.[4] Droplet coalescence is also an important aspect of many technologically important processes such as the growth of droplets in vapor deposited films[1,5], phase separation processes[6-8], combustion[8,9,10], the growth of bubbles in fluidized beds[11,12] and a variety of food products. Here I will describe some recent work in which simple models have been developed in an attempt to obtain a better understanding of droplet growth, deposition and coalescence. Much of this work has been carried out in collaboration with Fereydoon Family (Emory University) and the work on deposition with sliding was carried out with Zheming Cheng and Sidney Redner (Boston University).

Since we were interested primarily in universal scaling behavior associated with droplet coalescence processes, the physical and chemical details associated with such processes have been ignored. In all of these models we assume that the droplets have a (hyper) spherical shape and that the droplet radius, r, is related to its size (mass or volume), s, by

$$r \sim s^{1/D} \quad (1)$$

Here D is the dimensionality of the droplet. Most of our simulations have been carried out using integer values for D, but in general D can be considered to be a continuous variable. In all of the models it is assumed that when two (or more) droplets come into contact with each other, they coalesce with conservation of both the total mass and their center of mass. For the coalescence of two droplets of radius r_1 and r_2 the radius of the combined droplet is given by

$$r = (r_1{}^D + r_2{}^D)^{1/D} \quad (2)$$

After a pair of droplets has been coalesced in this fashion, the environment of the combined droplet is examined to determine if overlap with additional droplets has occurred. If overlap with other droplets is found, the process is continued until a single isolated droplet has been formed. It is assumed that the droplet coalescence process is fast compared to other processes occurring in the system. Clearly, these simplifying assumptions are not appropriate in

Springer Proceedings in Physics, Vol. 52 **Dynamics and Patterns in Complex Fluids**
Editors: A. Onuki · K. Kawasaki

all cases. However, models incorporating these features appear to provide an adequate representation of the behavior of a wide range of real systems.

In the most simple models, new droplets are added randomly to the system and/or the existing droplets are allowed to grow. However, the droplets themselves do not move (except as a consequence of coalescence). In other models we have allowed the droplets to move via either random walk or linear (ballistic) trajectories. Results for both types of models are described here.

2. Droplet Growth, Deposition and Coalescence

In a variety of processes a distribution of droplet sizes evolves as a result of the continuous addition of material to a substrate. The formation of droplets during thin film deposition is a typical example. In some cases the formation of droplets occurs at specific nucleation sites and in this case the process can be considered in terms of a droplet growth and coalescence process.[13,14] In other situations the droplet nucleation process is homogeneous and/or the objects being deposited are themselves macroscopic droplets. In this case a droplet deposition and coalescence model is more appropriate. These processes are closely related. Simple models describing them are presented in this section.

2.1 Droplet Growth and Coalescence

To represent the growth and coalescence of droplets in a system in which the droplets are formed by a heterogeneous nucleation process, we start by randomly placing N_o droplets of radius r_o in a system of size L^d (with periodic boundary conditions) avoiding overlap between the droplets. Here d is the dimensionality of the substrate. In all of the simulations described here the substrate is a continuous Euclidean space with a dimensionality (d) of 1, 2 or 3; however, similar simulations have been carried out using fractal substrates[15] embedded in a Euclidean space.[16] The radii of the droplets are assumed to grow algebraically according to the growth law

$$dr/dt \sim r^{\omega} \tag{3}$$

and the individual droplets grow according to equation (3) until they contact and coalesce. In practice, the radii of the droplets are increased in small steps according to equation (3) using the formula

$$r' = (r^{\xi} + \delta\xi)^{1/\xi} \tag{4}$$

where r' is the new radius, $\xi = 1-\omega$ and δ is a small number. The parameter δ is decreased until an increase in δ by a factor of at least 2 leads to changes in the simulation results that are smaller than the statistical uncertainties.

Figures 1 and 2 show the results of simulations carried out using two dimensional (d = 2) substrates. To obtain Figure 1 the parameters D = 2, ω = 1/2 and N_o = 20,000 were used. Similarly, for Figure 2 D = 3, ω = 0 and N_o = 20,000. In both cases the simulations were carried out using droplets with an initial diameter (d_o) of 1.5 (r_o = 0.75) on a substrate of area 512x512. In these simulations it was assumed that the centers of the droplets lie in the plane. Very similar results were obtained for spherical droplets lying on the plane. For the case D = d (Figure 1) a broad distribution of droplet sizes evolves which in the asymptotic limit ($t \to \infty$, $L = \infty$) has a power law form. For D<d (Figure 2) the distribution of cluster sizes is quite narrow and can be characterized by the "mean" cluster size S defined as

$$S(t) = \sum_{s=1}^{\infty} s^2 N_s(t) / \sum_{s=1}^{\infty} s N_s(t) \tag{5}$$

where $N_s(t)$ is the number of droplets of size s at time t.

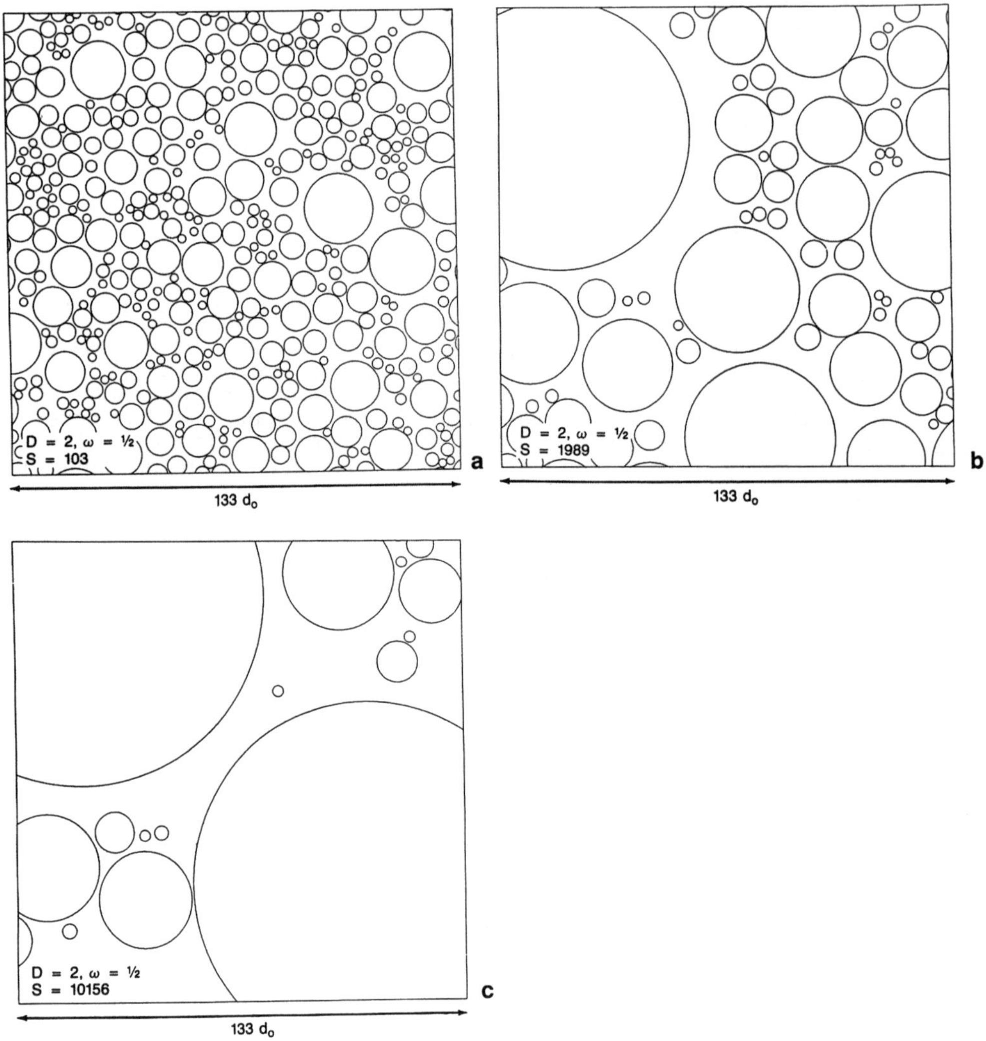

Figure 1

Three stages in a simulation in which two dimensional droplets grow and coalesce on a two dimensional substrate with a growth exponent (ω) of 1/2. Figures 1a-1c show a 200x200 ($133\ d_0 \times 133\ d_0$) region in a simulation carried out on a substrate of size 512x512 at three different stages at which the mean droplet size (S) has reached values of 103, 1989 and 10156 respectively.

Figure 2 shows how the system observed at different times can be rescaled to give a time independent pattern by contracting the system by a factor proportional to $S^{1/D}$. This means that the system can be described in terms of a characteristic length ($\simeq S^{1/D}$) and motivates a scaling model for the simulation results. Simulations carried out with various combinations of D, d and ω indicate that the mean droplet size S(t) and the number of droplets, N(t), both exhibit an algebraic time dependence

$$S(t) \sim t^z \qquad (6)$$

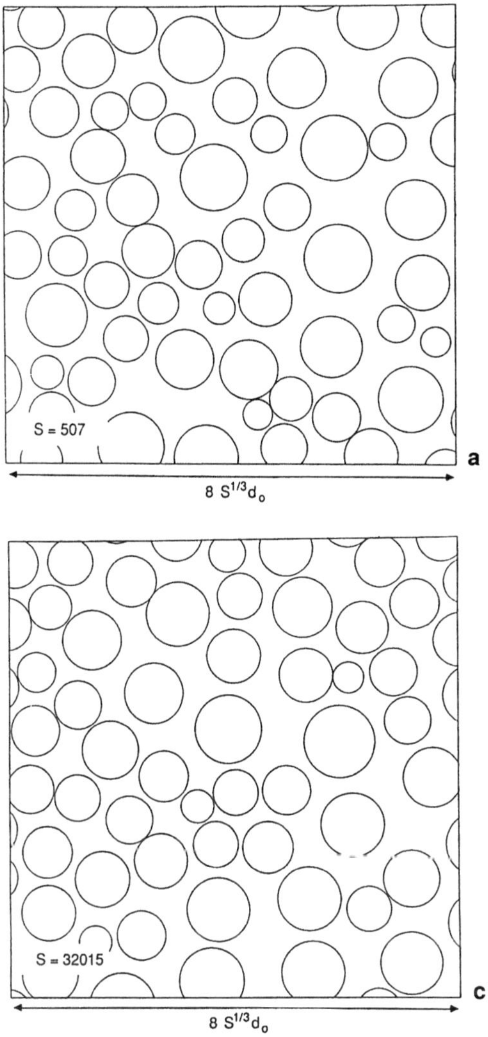

Figure 2

Results obtained from a simulation carried out using the parameters $D = 3$, $d = 2$, $\omega = 0$, $N_0 = 20{,}000$, $r_0 = 0.75$ and $L = 512$. Here three stages are shown on scales proportional to $S^{1/3}$ $(S^{1/D})$.

$$N(t) \sim t^{-z'} \tag{7}$$

with $z \simeq D/(1-\omega)$ and $z' \simeq d/(1-\omega)$.

Simulation results such as those shown in Figure 2 suggest (strongly) that for $D>d$ and $\omega<1$ the system has only one characteristic length scale $(R(t))$ whose growth is described by equation (3) with or without coalescence. Since $R(t) \sim t^{1/(1-\omega)}$ it follows immediately that $z = D/(1-\omega)$ and $z' = d/(1-\omega)$ in accord with the simulation results.

For the distribution of droplet sizes it is natural to propose the scaling form

$$N_s(t) \sim s^{-\theta} f(s/S(t)) \tag{8}$$

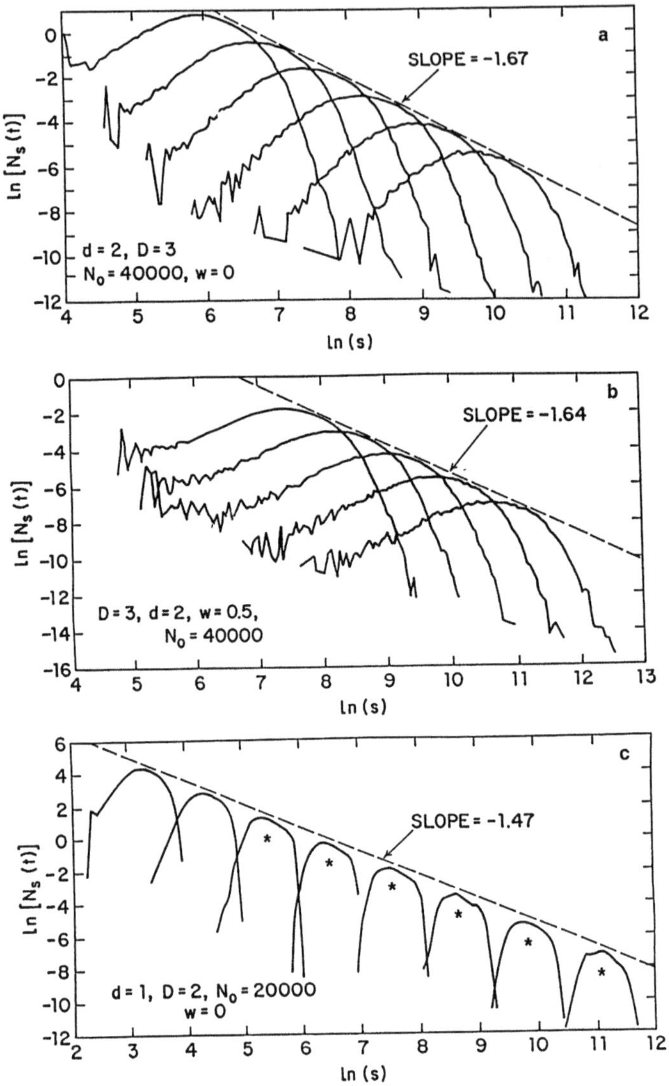

Figure 3

Droplet size distributions obtained from three different models for droplet growth and coalescence. (a) and (b) show results from simulations with D = 3 and d = 2 with growth exponents ω of 0 and 0.5, respectively. (c) shows the distribution for simulations with D = 2 and d = 1 with ω = 0.

that has proven to be appropriate for a broad range of aggregation processes[17-19]. For a system in which the total mass, M, or density, ρ, increases as t^y we find that

$$M = \Sigma s N_s(t) \sim \int ds\; s^{1-\theta} f(s/S(t)) \sim S^{2-\theta} \int dx\; x^{1-\theta} f(x) \sim t^{z(2-\theta)} \qquad (9)$$

so that

$$z(2-\theta) = y \qquad (10)$$

or

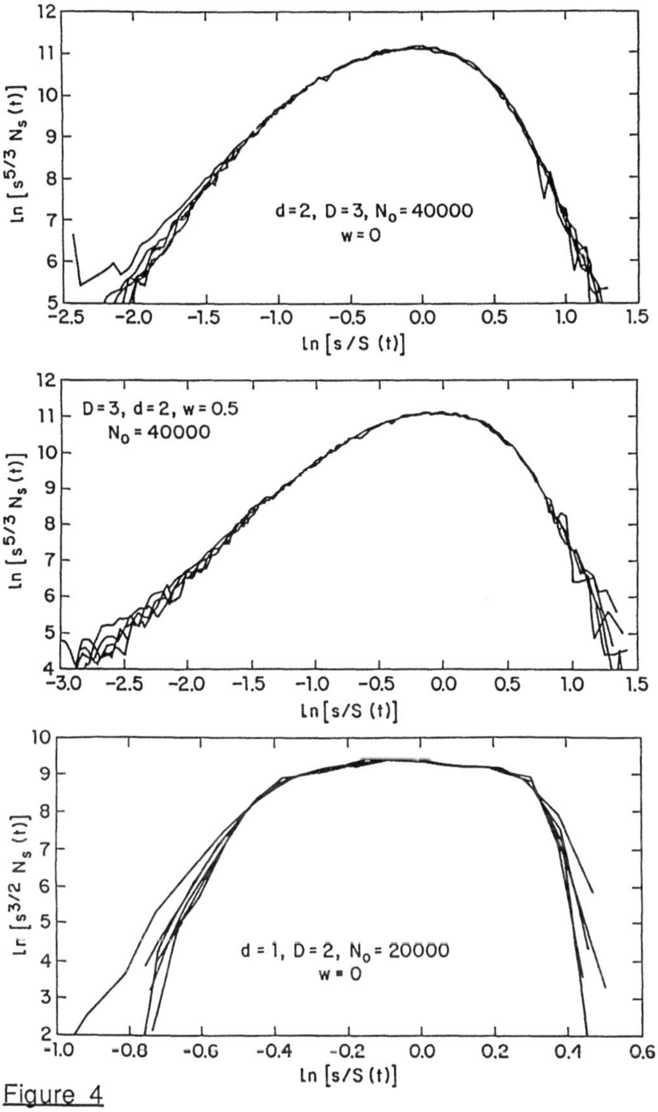

Figure 4

Scaling of the droplet size distributions shown in Fig. 3. The scaling form given in equation (5) was used with a scaling exponent θ of (D + d)/D (equation (26)).

$$\theta = (2z-y)/z \tag{11}$$

For mass conserving systems[17-19] y = 0 and θ = 2. In the present case y = z-z' and the exponent θ in equation (8) is given by

$$\theta = (z + z')/z = 1 + d/D \tag{12}$$

Figures 3 and 4 show some tests of this simple scaling theory. It is apparent that both the scaling form given in equation (8) and the values for the exponent θ given by equation (14) are well supported by simulation results.

2.2 Droplet Deposition and Coalescence

In many deposition processes material is added randomly to a surface or substrate in the form of small pieces with a relatively narrow size distribution. These "pieces" may be of molecular or larger sizes and if the substrate is held at a temperature near to or above the melting point of the deposited material, liquid or liquid-like droplets may be formed. Such processes may be represented by a simple model in which D dimensional droplets of equal size (radius r_o) are added, one at a time, to a d dimensional substrate. Figure 5 shows the results obtained by one such simulation with the parameters D = 3 and d = 2. In this simulation all of the added droplets had a radius of 0.75 and a substrate of size 512x512 was used (with periodic boundary conditions). At late times (Figure 5d) the system contains two distinct populations of droplets. A narrow distribution of large droplets which resembles that associated with the previous model for growth and coalescence and a broader distribution of small droplets. The scaling of the large droplets can be demonstrated in a manner similar to that used in Figure 1. Since a characteristic length R(t) is associated with the large droplets we assume that

$$S(t) \sim R(t)^D \tag{13}$$

and

$$N'(t) \sim R(t)^{-d} \tag{14}$$

where N'(t) is the number of "large" droplets. Since mass is added at a constant rate and almost all the mass is contained in the large droplets, we find that

$$M(t) \sim S(t)N'(t) \sim R(t)^{D-d} \sim t^1 \tag{15}$$

so that

$$S(t) \sim t^{D/(D-d)} \tag{16}$$

and

$$N'(t) \sim t^{-d/(D-d)} \tag{17}$$

The values of D/(D-d) and d/(D-d) can also be obtained by using the results from the droplet growth and coalescence model noting that the large droplets evolve by such a growth and coalescence process. In the limit in which the ratio between the radii of the large droplets (R) and the added droplets (r) diverges, the rate at which mass is added directly to the large droplets is proportional to R^d so that

$$\frac{dm}{dt} \sim R^d \tag{18}$$

where m is the droplet mass and

$$\frac{dR}{dt} = \frac{dR}{dm}\frac{dm}{dt} = R^{(1+d-D)} \tag{19}$$

so that the growth exponent (ω in equation (3)) is equal to 1+d-D. From this it follows that z = D/(D-d) and the exponent z" describing the algebraic decrease of N'(t) with increasing time is given by z" = d/(D-d) in accordance with equations (16) and (17). Similarly, the distribution of the large droplets can be described by the scaling form given in equation (8) with θ = 1 + d/D.

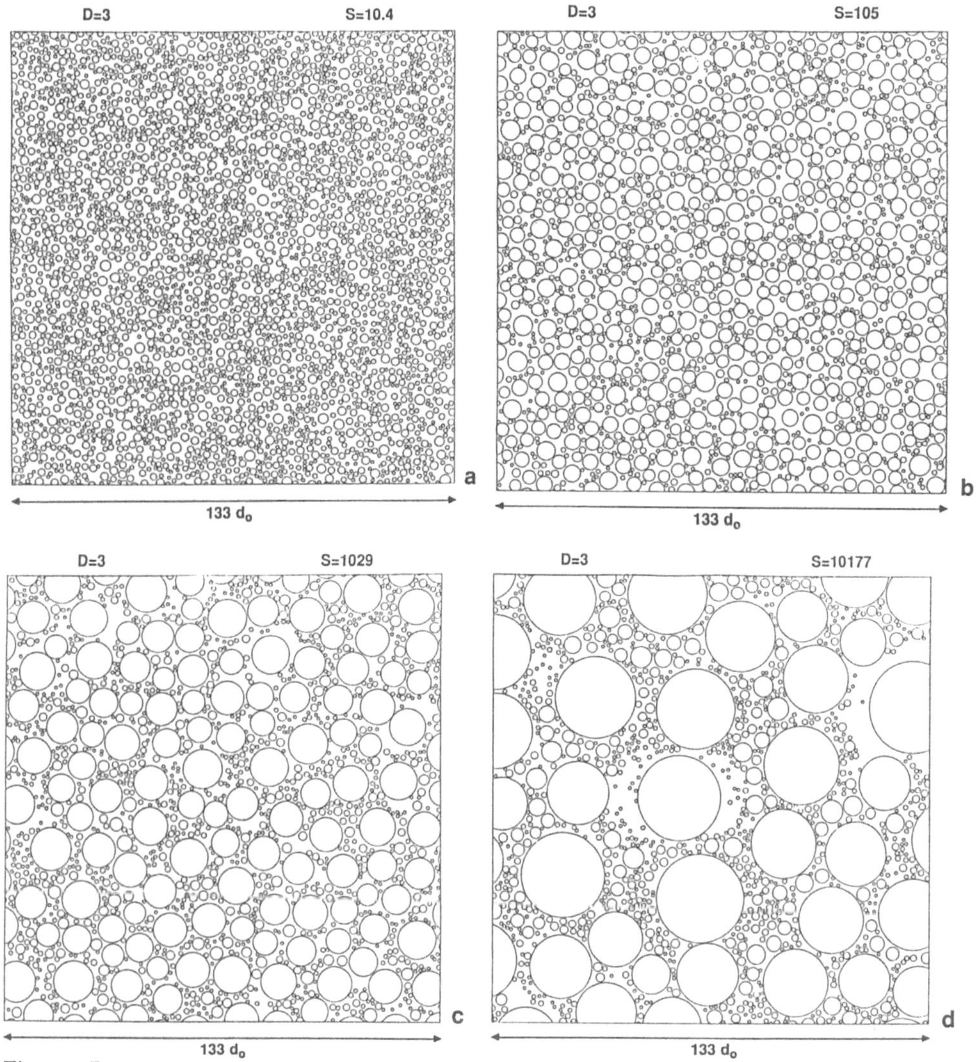

Figure 5

Four stages in a simulation in which three-dimensional droplets were randomly deposited onto a two-dimensional surface. Figures 5a-5d show 200x200 areas taken from simulations carried out on a 512x512 substrate (with periodic boundary conditions) at the instant at which the mean droplet size S(t) reached values of 10.4, 105, 1029 and 10177, respectively. The droplet size distribution consists of a monodispersed distribution of larger droplets superimposed on a highly polydisperse distribution of smaller droplets in the backbground. We find bimodal distribution of droplets in d = 1, 2 and 3 for the cases where D>d.

Figure 6 shows the scaling of the droplet size distribution for the physically most important case D = 3 and d = 2. It appears from Figure 6 that not only the distribution of large droplets but the entire droplet size distribution can be expresed in terms of the simple scaling form given in equation (8). The scaling function f(x) in equation (8) can now be written in the form

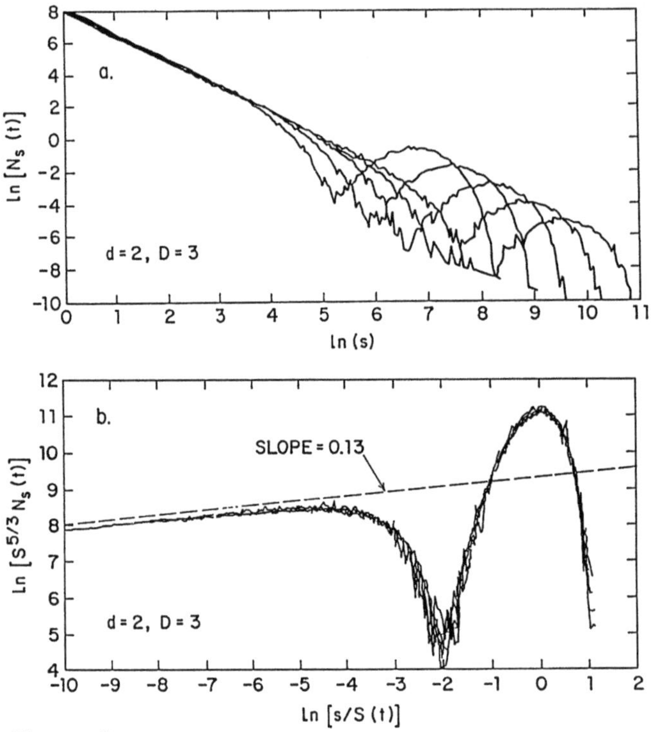

Figure 6

Scaling of the droplet size distributions for the D = 3, d = 2 case. (a) shows the droplet size distribution at five late stages during the simulation taken from the simulations used to obtain Figure 6. (b) shows how these size distributions can be scaled onto a common scaling function using the scaling form given in equation (5) with a scaling exponent θ of 5/3 (1+d/D). The nonzero slope for small values of the argument of the scaling function indicates that $\theta-\tau>0$.

$$f(x) = x^{(\theta-\tau)}\, g(x) + h(x) \tag{20}$$

The function g(x) that describes the distribution of small droplets has a constant value for small x and decreases faster than any power of x for large x. The function h(x) that describes the distribution of large droplets has a peak near x = 1 and decays rapidly for both large and small values of x. A comparison of the function h(x) obtained from simulations of droplet deposition and coalescence is compared with the function f(x) for droplet growth and coalescence in Figure 7. In both simulations D = 3 and d = 2. A value of 0 (1+d-D) was used for the exponent ω in the growth and coalescence simulations. Both simulations were carried out using an initial droplet radius (r_0) of 0.75 on substrates of size 512x512.

Since almost all of the droplets are "small", the total number of droplets N(t) is given by

$$N(t) \simeq \sum_{s=1}^{s^*} s^{-\tau} \tag{21}$$

where s* is the cut-off in the distribution of small droplets. For $\tau>1$ the sum

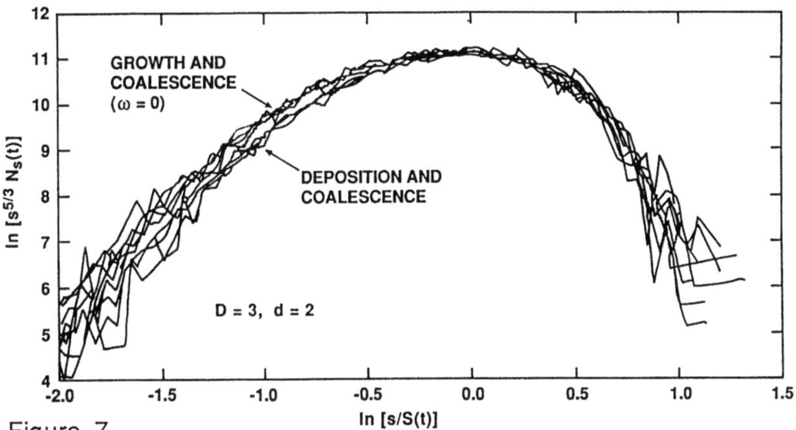

Figure 7

Scaling of the large droplet part of the size distributions in the deposition and coalescence and growth and coalescence models. The similarity of the two distributions is quite evident and points to the conclusion that the large droplets part of the distribution in the homogeneous nucleation model arise from growth and coalescence of similar size droplets.

in equation (21) is dominated by the very small droplets and $N(t) \simeq N_1$ (the number of droplets of size s = 1). The quantity N_1 can be estimated by recognizing that both the large and small droplets occupy an essentially fixed fraction of the total volume (area length) in the system. The total "volume" of small droplets is given by

$$V(t) = N_1 \sum_{s=1}^{s^*} s^{-\tau} s^{d/D} \tag{22}$$

or

$$V(t) \sim N_1 \int_0^{s^*} ds\, s^{-\tau} s^{d/D} \tag{23}$$

Since $\theta = 1 + d/D$ we find that

$$V(t) \sim N_1 s^{*(\theta-\tau)} \tag{24}$$

if $\theta > \tau$. Since the cut-off s^* in the distribution grows in the same way as S, the mean droplet size or the characteristic size of the large droplets in equation (24) can be replaced by

$$V(t) \sim N_1 t^{z(\theta-\tau)} \tag{25}$$

and if τ lies in the range $1<\tau<\theta$ then

$$N(t) \simeq N_1 \simeq t^{-z(\theta-\tau)} \tag{26}$$

or

$$z' = z(\theta-\tau) \tag{27}$$

Equation (27) can also be obtained from more general scaling arguments.[20-23] The exponent z' is quite small and as a consequence is difficult to measure accurately. However, for those cases where z' can be measured, equation (27) appears to be satisfied.

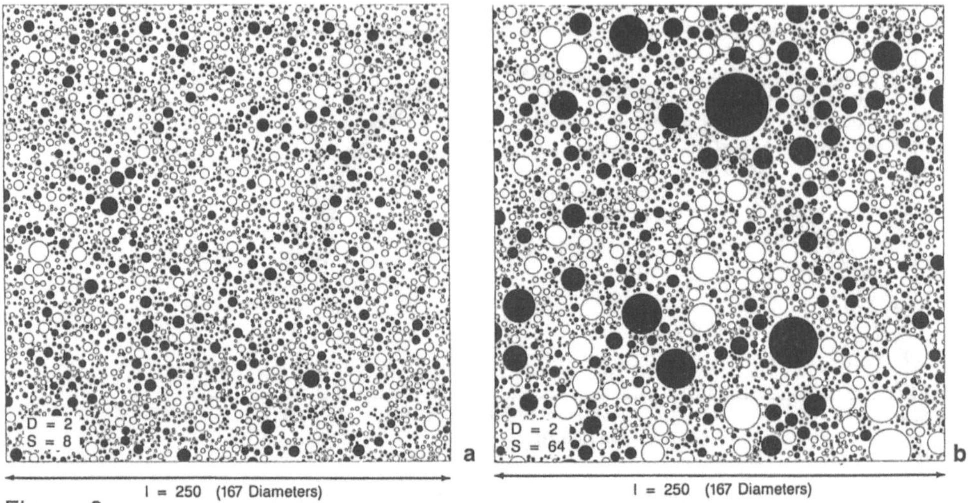

Figure 8

Two stages in a simulation of the deposition, coalescence and annihilation of droplets on a surface. In this case equal sized two dimensional droplets with randomly selected positive and negative masses were deposited randomly on a two dimensional surface. Figure 8a shows part of the system at a stage at which a mean droplet size (S) of 8 had been reached. Figure 8b shows the same region at the stage where S has grown to a value of 64.

Simulations have also been carried out in which droplets with a mass of $\pm m_0$ are added randomly to a d dimensional substrate. In this case complete or partial annihilation occurs when masses of opposite signs contact each other. Again, it appears that the evolution of the system can be described in terms of a simple scaling picture. There is a characteristic length R associated with the large droplets. A large droplet of size s can be formed only if s^2 "monomer" droplets of radius r_0 have been added to a region with a "volume" of $s^{d/D}$. This will require a time t proportional to $s^{2-d/D}$ so that the exponent z describing the growth of the mean droplet size is given by

$$z = D/(2D-d) \qquad (28)$$

Similarly, the droplet size distribution can be described by the scaling form in equation (8). The exponent θ is given by equation (12). Figure 8 shows the results of a simulation carried out using this model with the parameters D = d = 2. In this simulation the position of the center of a droplet formed by adding n smaller droplets with masses of opposite sign is given by

$$\mathbf{r} = \sum_{i=1}^{n} |m_i|\mathbf{r}_i \Big/ \sum_{i=1}^{n} |m_i| \qquad (29)$$

Simulations carried out using this model with various combinations of d and D (with $D \geq d$) give results that are consistent with the simple scaling arguments outlined above. For example, Figure 9 shows the time dependence of S and N obtained from a simulation with the parameters D = 3 and d = 2. The effective value of z (0.73) agrees quite well with that obtained from equation (28) (z = 3/4). Similarly, Figure 10a shows the droplet size distribution (distribution of |m|) at seven stages during these simulations and Figure 10b shows that these distributions can be scaled onto a common curve using the scaling form given in equation (8) with the theoretical value for θ (θ = (D+d)/D = 5/3).

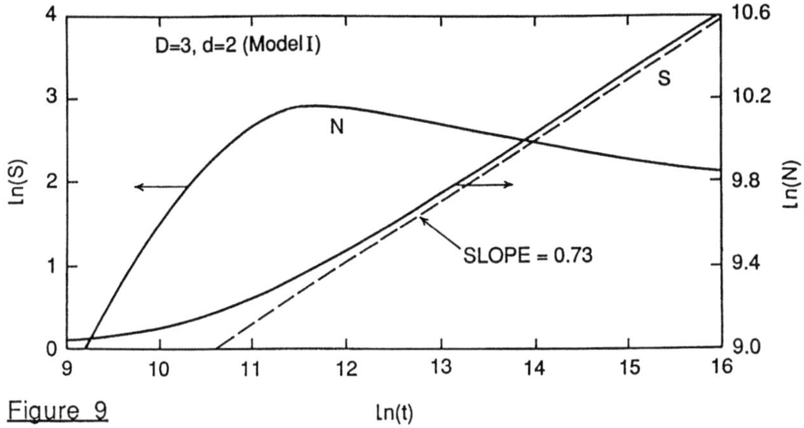

Figure 9

The growth of the mean droplet size (S) and the decay of the number of droplets (N) obtained from simulations carried out using the droplet deposition, coalescence and annihilation model. In this simulation D = 3, d = 2 and the center of mass of coalesced droplets is given by equation (29).

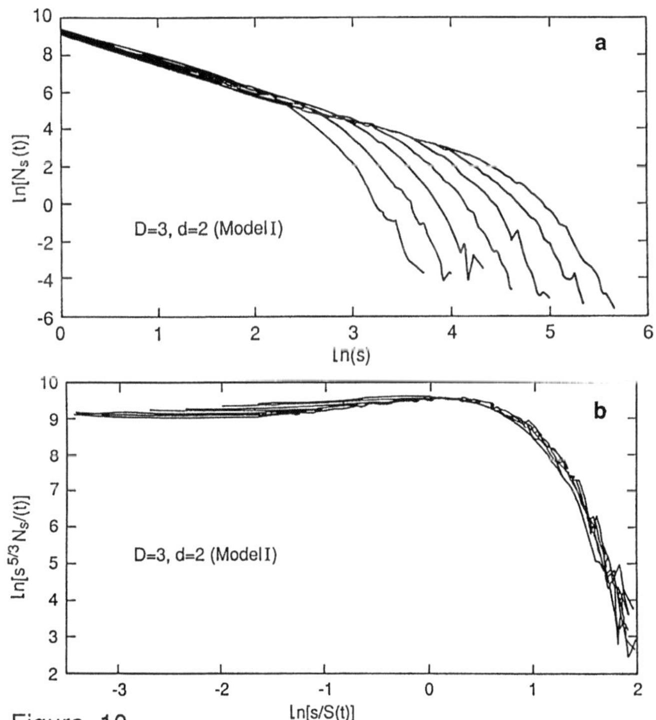

Figure 10

Droplet size distributions obtained from the deposition coalescence and annihilation model used to obtain the results shown in Figure 9. Figure 10a shows the size distribution ($N_S(t)$) at seven stages during the simulations and Figure 10b shows how these curves can be scaled using the scaling form given in equation (8) and the theoretical value for the scaling exponent θ ($\theta = (D+d)/D$).

Although these do not appear to be physically realistic situations in which positive and negative masses are added to the substrate, this model may be appropriate for the deposition of two chemically different materials that react to give a product that is removed by evaporation, sublimation, absorption by the substrate, etc.

3. Models with Droplet Motion

In many circumstances droplets can move as a consequence of thermal fluctuations, external forces or droplet-droplet interactions. Simple scaling behavior, similar to that described above, is frequently found in these systems as well.

3.1 Brownian Coalescence

To simulate the coalescence of droplets whose motion is a consequence of Brownian motion alone, the droplet trajectories can be represented by random walks in which the step sizes δ are much smaller than the droplet diameters. In this model it is assumed that the droplet diffusion coefficients $\mathcal{D}(s)$ have an algebraic dependence on their sizes

$$\mathcal{D}(s) \sim s^{\gamma} \tag{30}$$

This is implemented in the model by selecting droplets randomly with a probability proportional to s^{γ} and moving them by a constant amount (δ) in a random direction. Figure 11 shows some results from a simulation in which 72,000 (N_0) two dimensional droplets with a diameter d_0 of 1.5 were initially placed at random on a two dimensional surface with an area of 512x512. The step size (δ) was adjusted to 1/8 of the diameter of the smallest droplet throughout the simulation and the exponent γ in equation (30) was set to a value of -1/2. Figure 12 shows the time dependence of the mean droplet size (S(t)) and the number of droplets (N(t)) obtained from 7 simulations carried out using the same parameters that were used to generate Figure 11. The results in this figure indicate that the exponents z and z' (equations (6) and (7)) both have values close to 2/3. For the case D = d the droplet diameters grow in the same way as the distances between the droplets so that patterns such as those shown in Figure 11 can be rescaled in the way that is illustrated in the droplet growth and coalescence model (Figure 2). Consequently, the system can be characterized by a single length (R). Since the coalescence process is diffusion-limited, the droplet collision frequency υ is given by

$$\upsilon \sim R^{-2}R^{\gamma D}N \sim R^{(d(\gamma-1)-2)} \tag{31}$$

or

$$\frac{dN}{dt} \sim -N^{(-\gamma+1+2/d)} \tag{32}$$

Consequently, we find that $z' = 1/((2/d)-\gamma)$ and for d = 2 the exponents z and z' are both equal to $1/(1-\gamma)$ in accord with the results shown in Figure 12 and other simulations carried out with the parameters D = d = 2 and various values for γ and N_0. For the case D = d = 3, equation (32) leads to the result $z = z' = 1/(1-1/D-\gamma)$ that is also in good agreement with simulations carried out using droplets with an initial diameter of 1.8 and system sizes of 64x64x64 or 128x128x128 (with periodic boundary conditions) for several values of γ in the range $-1 \leq \gamma \leq 0$.

If the dimensionalities of the droplets and substrates are not equal, the length scales associated with the droplet sizes and droplet separations are not equal and the behavior becomes more complex.

For the case d = 2 and $D \rightarrow \infty$ (particle coalescence), it is now well known that the asymptotic time dependence of N(t) and S(t) for $\gamma = 0$ is given by[24]

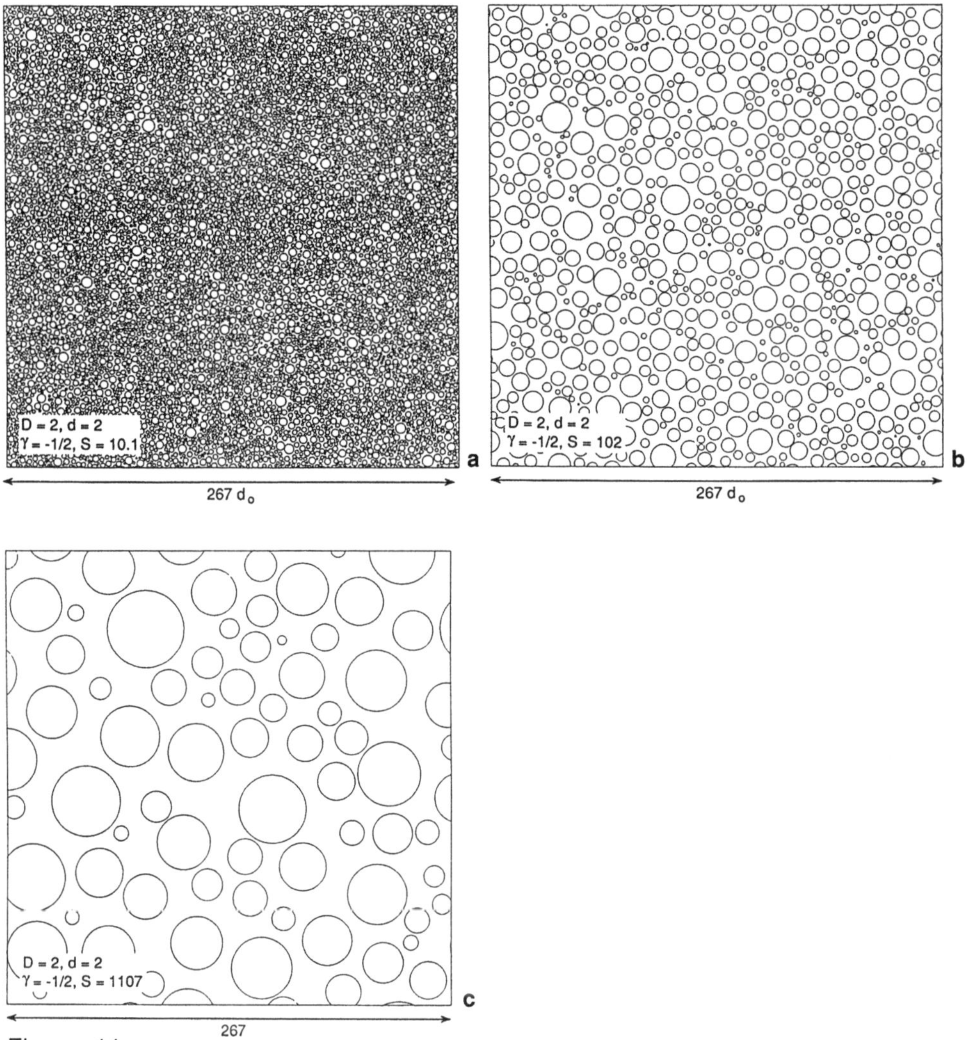

Figure 11

Three stages in a simulation of the Brownian coalescence of two dimensional droplets on a two dimensional substrate with a value of -1/2 for the diffusion coefficient exponent γ. Figures 11a, 11b and 11c show the system at the stages where the mean cluster size S has reached values of 10.1, 102 and 1107 respectively.

$$S(t) \sim t/\ell n(t) \qquad (33)$$

$$N(t) \sim \ell n(t)/t \qquad (34)$$

For the more general case (finite D not equal to d) the length associated with the droplet sizes is given by

$$R_S \sim N^{-1/D} \quad \text{or } R_S \simeq (\rho_n(t)/\rho_n(0))^{-1/D} \sim \rho_n^{-1/D} \qquad (35)$$

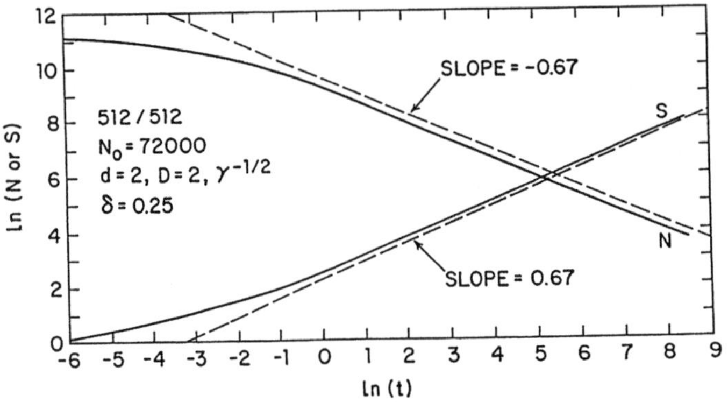

Figure 12
Time dependence of S and N obtained from a simulation of Brownian coalescence with a two dimensional substrate and two dimensional droplets.

where ρ_n is the number density of the droplets while the length associated with the distances between them is given by

$$R_d \sim N^{-1/d},\ R_d \simeq \rho_n^{-1/d} \tag{36}$$

Consequently, the fraction f of the surface covered by droplets is given by

$$f \sim (R_s/R_d)^d \sim N^{(1-d/D)},\ f \simeq \rho_n^{(1-d/D)} \tag{37}$$

A consequence of this is that the collision frequency for the case d = 2 is given by

$$\upsilon \sim R_s^{-2}\ R_s^{\gamma D}\ \rho_n^{(1-d/D)}\ \rho_n/\ell n(\rho_n^{-1}) \tag{38}$$

Equation (38) was obtained by exploiting the self-similarity[15] of the random walk trajectory describing the relative motion of droplet pairs. The random walk trajectory is coarse-grained into regions or blobs with a characteristic size equal to R_s. The first two terms on the right hand side of equation (38) ($R_s^{-2}R_s^{\gamma D}$) come from the time required for the droplets to move by a distance R_s. The factor of $(\ell n\rho_n^{-1})^{-1}\rho_n^{(1-d/D)}$ comes from the observation that the number of steps required for the coarse-grained random walk to enter f^{-1} new blobs is asymptotically proportional to $f^{-1}\ell n(f^{-1})$ so that the collision frequency is proportional to $f/\ell n(f^{-1})$. Consequently, equation (38) can be written as

$$\frac{d\rho_n}{dt} \simeq (\ell n(\rho_n))^{-1}\rho_n^{(2/D-\gamma+2-d/D)} \tag{39}$$

so that the asymptotic time dependence of N(t) is given by

$$N(t) \sim (t/\ell n(t))^{-(1/(1-\gamma))} \tag{40}$$

The simulation results for various combinations of N_0, D and γ are consistent with equation (40). For example, Figure 13 shows the dependence of N(t) and S(t) on $t/\ell n(t)$ for 3 simulations carried out with D>2 on two dimensional substrates.

For Brownian coalescence in three dimensional systems the exponents z and z' are given by

$$z = z' = 1/(2/3 - \gamma) \tag{41}$$

for the case D = d (from equation (32)).

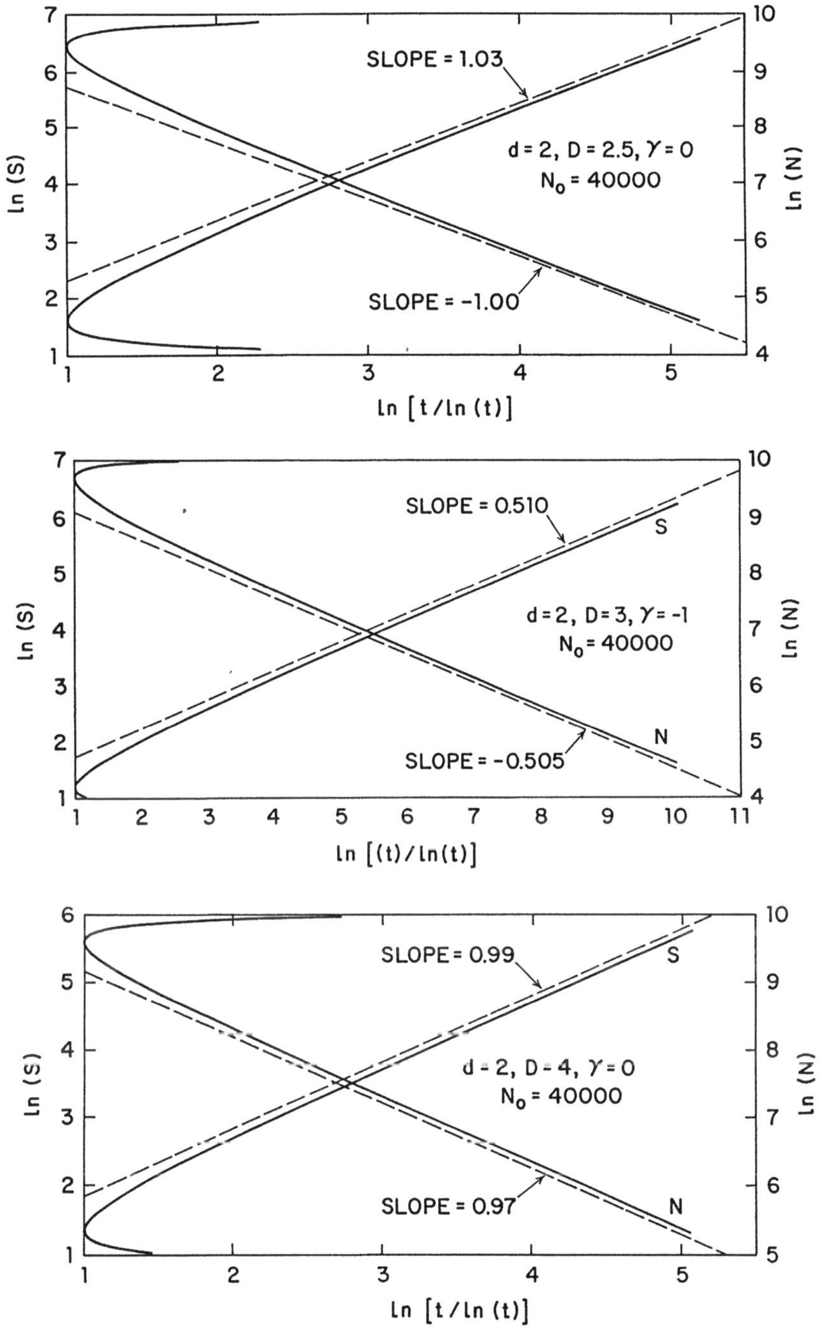

Figure 13

The time dependence of S and N obtained from simulations of the coalescence of droplets with dimensionalities (D/>2) on a two dimensional substrate. These results are in good agreement with equation (40).

In the mean field limit the Brownian coalescence process can be described by a Smoluchowski equation[25,26] with the reaction kernel

$$K(i,j) = (\mathcal{D}_i + \mathcal{D}_j)\ (r_i^{d-2} + r_j^{d-2}) \tag{42}$$

for d>2. Here K(i,j) is the rate constant corresponding to the coalescence of droplets of size i and j. The homogeneity index λ for this reaction kernel is given by

$$K(i,i) \sim i^{\lambda} \tag{43}$$

and

$$\lambda = \gamma + (d-2)/D \tag{44}$$

Consequently, we expect that the exponents z and z' will both have the value of $1/(1-\lambda)$ or

$$z = z' = D/(D-D\gamma-d+2) \tag{45}$$

For the case D = d this becomes

$$z = z' = 1/(2/d-\gamma) \tag{46}$$

in agreement with equation (32). Simulations carried out in three dimensions using droplets with an initial radius (r_o) of 0.9 in systems of size 64x64x64 and 128x128x128 give values for z and z' in good agreement with equation (45) for simulations carried out with D = 3 and 4 with several values for γ.

In all cases the asymptotic distribution of droplet sizes can be described in terms of the scaling form given in equation (8) with $\theta = 2$. This value for θ is expected since the total mass is conserved in the Brownian coalescence simulations. Most of the Brownian coalescence simulations have been carried out with D>d. In the case D<d a process analogous to gellation occurs and the mean droplet size S diverges in a finite time.

3.2 A Simple Avalanche Model for Deposition with Sliding

In this model[25] we imagine that particles are deposited continuously onto an inclined substrate. When a droplet reaches a size s_c as a result of the deposition and coalescence process described above, the droplet slides "down hill" collecting with it all other droplets that it contacts. Figures 14 and 15 show the results of simulations carried out using this model for the case of deposition of D (2, 3 and 4) dimensional droplets on a two dimensional substrate. Simulations have also been carried out using lattice models for both one and two dimensional substrates. In these models an "avalanche" occurs where the height h associated with one of the lattice sites reaches a pre-assigned threshold value h_c. Figure 16 illustrates the two dimensional version of this model with $h_c = 3$. In these models the avalanche is assumed to be complete on the time scale that is short compared with that needed to add one droplet or site to the system and the avalanche ceases only when it reaches the (lower) boundary and its mass is removed. Using these models both the initiation of the first avalanche and the properties of the fluctuating steady state have been explored. In all of the models periodic boundary conditions are used in the lateral direction. Here attention is focussed on some aspects of the more realistic off-lattice models with two dimensional substrates.

Using simple statistical arguments, it can be shown that the "time" required (on average) for h_c particles to land on the same lattice site in a lattice containing J sites is given by

$$\tau \sim J^{\alpha} \tag{47}$$

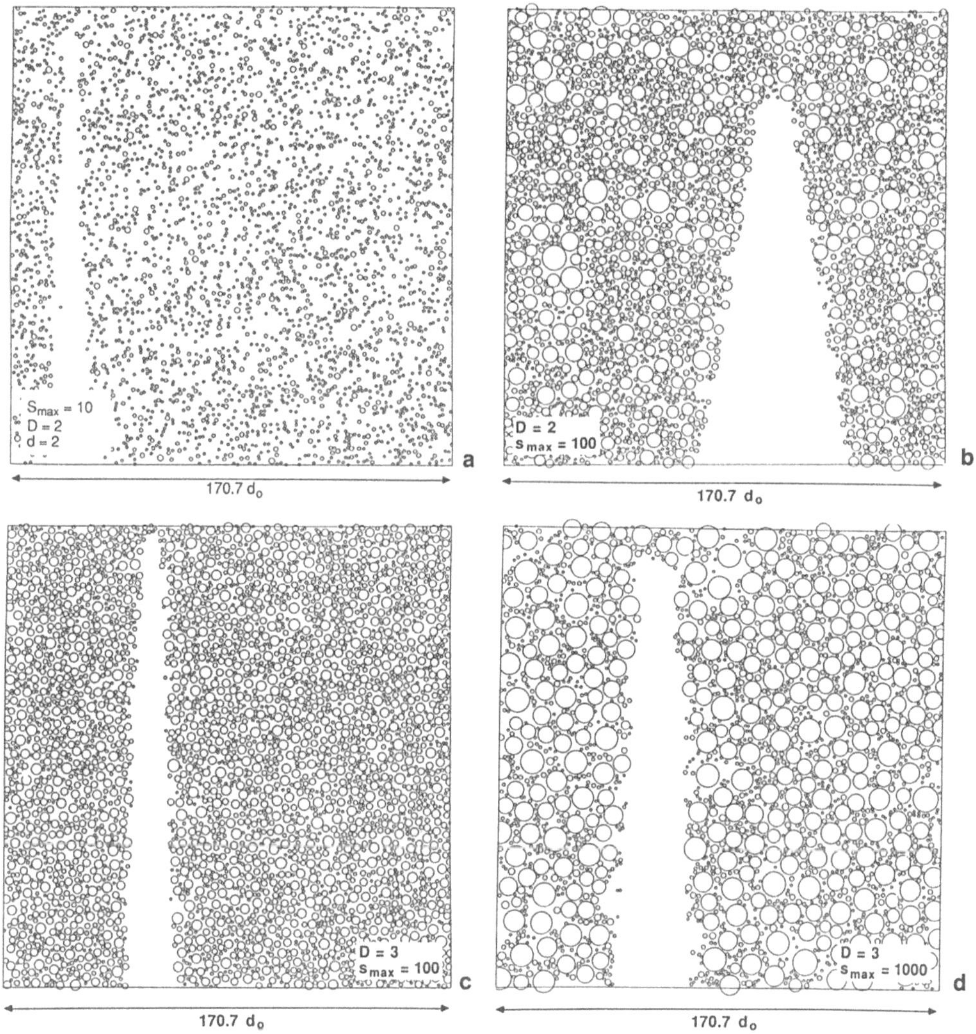

Figure 14

Results obtained from the two dimensional off-lattice model for droplet deposition, coalescence and sliding. Here several systems immediately after the first "avalanche" has occurred are shown. D is the dimensionality of the droplets and S_{max} is the maximum droplet size ($h_c = S_{max} + 1$).

where the exponent α is equal to $1\text{-}1/h_c$. For the off-lattice case an avalanche is initiated when a mass of $\simeq h_c$ has been deposited in a region of size $h_c^{d/D}$ and the effective number of lattice sites is given by

$$J' \simeq L^d h_c^{-(D/d)} \tag{48}$$

so that $\tau \sim L^{(1-1/h_c)d}$ where L is the system size (length). Both lattice and off-lattice simulations are in good agreement with this idea. In the case of a one dimensional system the mass M_1 removed by the first avalanche contains (on average) a substantial fraction of the total mass deposited so that we expect (and find) that

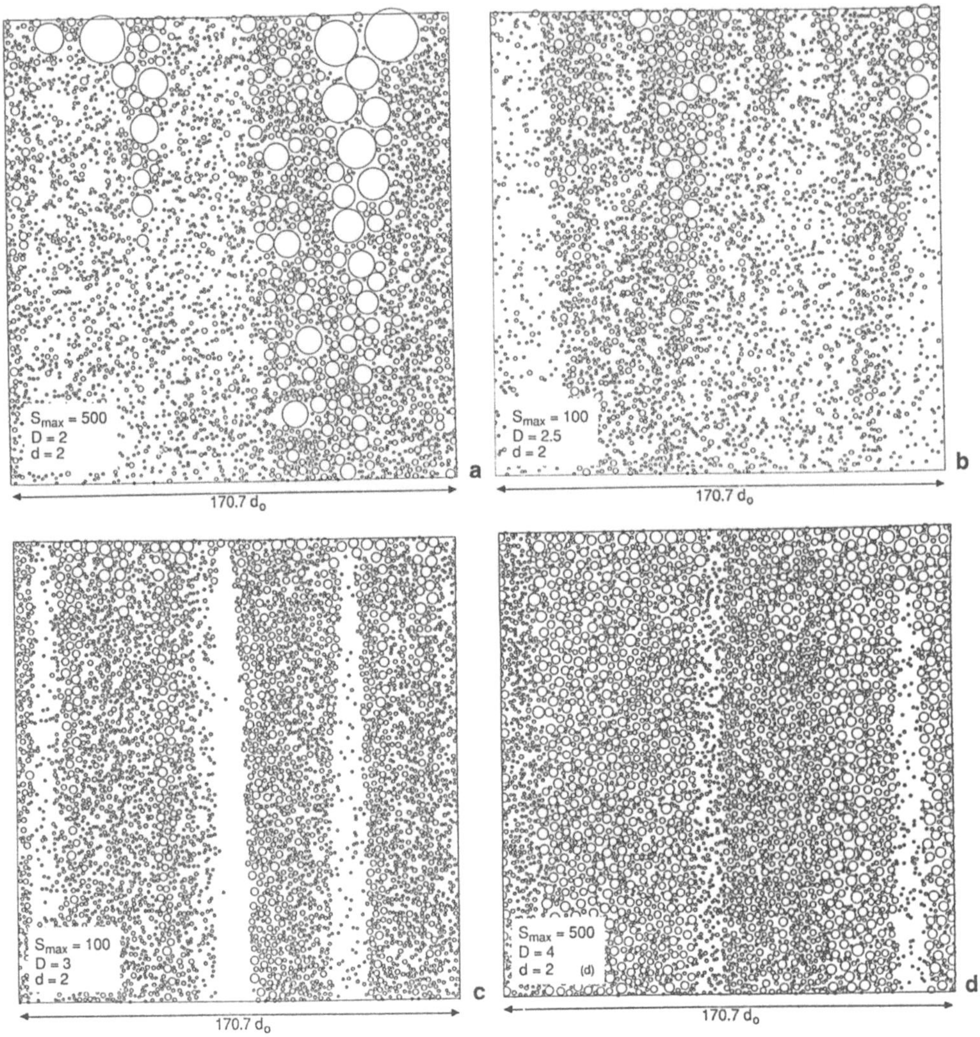

Figure 15

Typical steady state configurations obtained from the off-lattice model for droplet deposition, coalescence and sliding on a two-dimensional surface. In both Figures 14 and 15 the droplets slide in the direction from the top to the bottom of the figures.

$$M_1 \sim J^{\alpha'} \tag{49}$$

where $\alpha' = \alpha = 1-1/h_c$.

For the case of two dimensional droplets on a two dimensional substrate the first avalanche removes a wedge-shaped region from the substrate (see Figure 14b, for example). The average mass contained in this region depends on the wedge angle α. Since the first avalanche occurs when the density reaches a value of the order of L^{-2/h_c}, the angle α is given by $\tan(\alpha/2)$ α L^{-2/h_c} and the average mass removed is given by

$$M_1 \sim L^{2(1-1/h_c)}\, L^{-2/h_c} \tag{50}$$

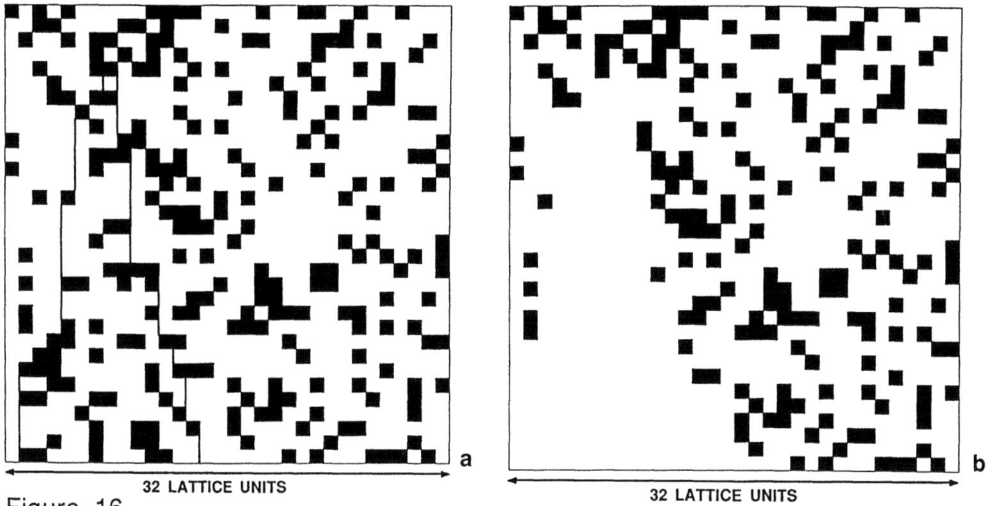

Figure 16

Illustration of the discrete version of the avalanche model on a 32x32 substrate for $h_c = 3$. The system is shown (a) just before and (b) just after the avalanche. For this system size, the wedge-shaped cleared region is apparent.

or

$$M_1 \sim L^{2(1-2/h_c)} \tag{51}$$

For the more general case of D dimensional droplets on a two dimensional substrate, M_1 is related to L by

$$M_1 \sim L^{D(1-2/h_c)/(D-1)} \tag{52}$$

The simulations are in good agreement with this prediction for large values of h_c. For smaller values of h_c the agreement is not so good. This is probably a consequence of finite size effects associated with the assumption that the mass was uniformly distributed over the substrate just prior to the first avalanche. This mean field or continuum picture becomes less realistic for small values of L and h_c.

The statistical properties of primary interest in this work were the distribution P(t) of times before the first avalanche and between avalanches in the steady state and the distribution R(m) of the masses removed by the avalanches. Figure 17 shows these quantities obtained for the first avalanche from two dimensional off-lattice simulations with D = 2 and d = 2. It is apparent from Figure 17a that for large values of h_c $R(M_1) \sim M_1^{-1/2}$ and for simulations with three dimensional droplets $R(M_1) \sim M_1^{-1/3}$ for $h_c >> 1$. One dimensional simulations lead to the result $R(M_1) \sim M_1^{0}$ for $h_c >> h$. These results are a simple consequence of the shapes of the regions swept out by the avalanches and the random position of the avalanche initiation event. For the general case of D dimensional droplets on a d dimensional substrate the mass removed by an avalanche starting a distance y from the "lower edge" of the substrate is given by

$$M_1(y) \sim y^{D/(D-d+1)} \tag{53}$$

in the continuum limit. Consequently, we have

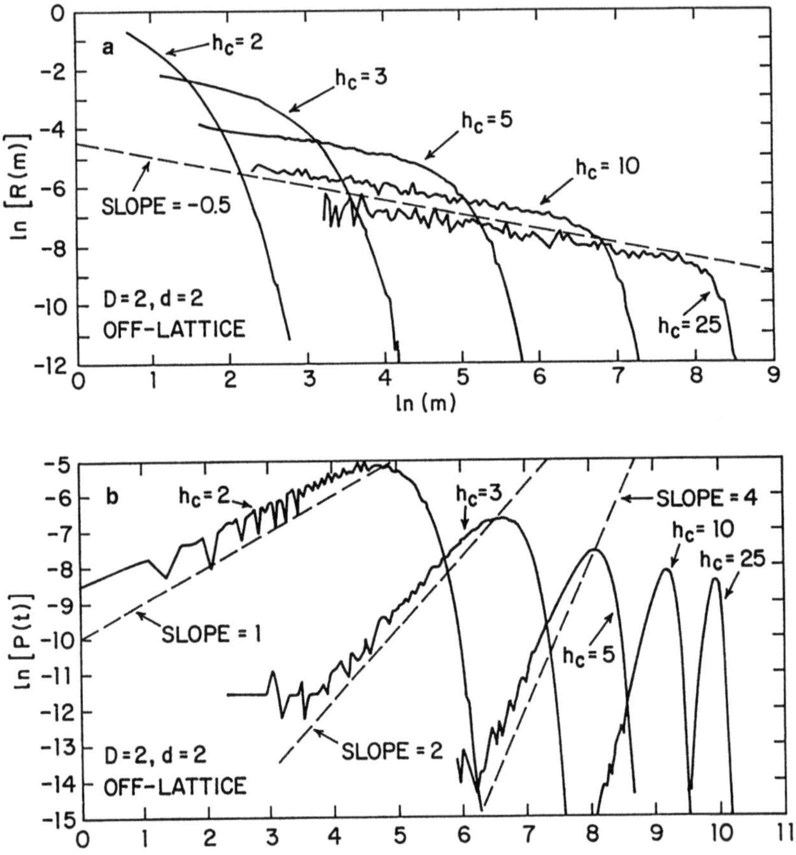

Figure 17

Distribution of (a) the mass removed by the first avalanche, R(m), and (b) the times until the first avalanche P(t) obtained from off-lattice simulations of two-dimensional (D = 2) droplets for several values of h_c.

$$\frac{dM_1(y)}{dy} = y^{(d-1)/(D-d+1)} \tag{54}$$

and

$$R(M_1) \sim M_1^{(d-1)/D} \tag{55}$$

Figure 17b shows that $P(t) \sim t^{h_c-1}$. This result is correct for all values of D and d and is a consequence of the uncorrelated nature of the random deposition process.[27]

These models for deposition, coalescence and sliding were motivated in part by the concept of self-organized criticality introduced by Bak et al.[28,29] However, the autocorrelation function for the mass flux leaving the system under steady state conditions obtained from both one dimensional and two dimensional simulations shows no evidence of power law behavior and the models described here do not appear to be examples of self-organized criticality. This may be a consequence of the instantaneous propagation of the avalanches without any dissipative processes.

4. Summary

Several simple models in which droplet coalescence plays a key role have been described. In all of these models the kinetics of the coalescence exhibits simple scaling behavior that can be related to the geometry of the evolving systems. The behavior of these systems can be understood in terms of simple scaling arguments and statistical theories. Although these models neglect many of the detailed physical and chemical processes occurring in real systems, they appear to be insensitive to model details and it seems probable that a broad variety of real systems will exhibit the same scaling behavior (belong to the same universality class). For example in both the droplet deposition growth and coalescence model and the Brownian coalescence the same scaling behavior is found for spherical droplets that just contact a two dimensional surface and spherical droplets whose centers lie in the plane.

The results obtained from the Brownian coalescence models suggest that the critical dimensionality for this process is 2. Similar results and conclusions have been obtained for diffusion-limited particle coalescence aggregation and annihilation models.[27,32-35] However, recent theoretical work by van Dongen[36] and simulation results[37] indicate that behavior of S(t) and N(t) alone does provide a completely reliable basis for determining if the system is above, at or below the critical dimension d_c (mean field theories may give the correct behavior for some quantities even though $d \leq d_c$).

A variety of other droplet coalescence models have been (or are being) explored. These include the coalescence of droplets that drift with a velocity V(s) that depends algebraically on droplet mass

$$V(s) \sim s^{\gamma} \quad (56)$$

in a manner similar to that describing the dependence of the diffusion coefficient $\mathcal{D}(s)$ on droplet size in the Brownian coalescence model. If point masses (with a distribution of these masses) are injected, one at a time at the end of a line, then asymptotic dependence of S and N are given by[38]

$$S(t) \sim t \quad (57)$$

and

$$N(t) \sim \ell n(t) \quad (58)$$

The mean particle mass at a distance ℓ from the injection point is given by

$$\bar{s}(\ell) \sim \ell^{1/(1+\gamma)} \quad (59)$$

The mean density at a distance ℓ from the injection point is given by

$$\bar{\rho}(\ell) \sim \ell^{\gamma(1+\gamma)} \quad (60)$$

and the mean distance $\bar{x}(t)$ moved by the particles at time t is given by

$$\bar{x}(t) \sim t^{(1+\gamma)} \quad (61)$$

The particle mass distribution $N_s(t)$ can be described by the scaling form

$$N_s(t) \sim s^{-1} f(s/S(t)) \quad (62)$$

These and other scaling relationships can be understood in terms of a simple scaling picture similar to those described above.

Similar results have also been obtained for the coalescence and drift of D dimensional droplets on/in d dimensional substrates for d = 2 and 3. Other models that have been investigated include non-uniform (spatially correlated) desposition and coalescence, continuous deposition, drift and coalescence of droplets on an "inclined" substrate and the drift and coalescence of droplets with an initial random distribution of droplet positions with a small distribution of droplet sizes. Most of these models appear to exhibit simple asymptotic scaling behavior but they have not yet all been analyzed in detail.

While analysis of simulation results using scaling theories is both simple to implement and gratifying in terms of the agreement between the measured and predicted exponents, understanding on a more detailed level is for the most part lacking. The form of the scaling functions describing droplet size distribution can, in principal, be calculated from Smoluchowski or other kinetic equations. However, analytical solutions are available for only a few simple kernels and the conditions under which mean field kinetic equations provide a satisfactory solution are not yet well understood.

5. References

1. D. Beysens and C. M. Knobler, Phys. Rev. Lett. 57, 1433 (1986).
2. J. L. Viovy, D. Beysens and C. M. Knobler, Phys. Rev. A 37, 4965 (1988).
3. D. Fritter, C. M. Knobler, D. Roux and D. Beysens, J. Stat. Phys.
4. B. Mason, "The Physics of Clouds", Oxford University Press, London (1957).
5. B. Lewis and J. C. Anderson, "Nucleation and Growth of Thin Films", Academic, New York (1987).
6. J. D. Gunton, M. San Miguel and P. S. Sahni in "Phase Transitions and Critical Phenomena", C. Domb and J. L Lebowitz, eds., Academic, London (1983), Vol. 8.
7. E. Siggia, Phys. Rev. A20, 595 (1979).
8. R. J. Quinn and A. F. Sarofim, Fuel 65, 40 (1986).
9. A. R. Kerstein, "Combustion Flame" (1989), to be published.
10. S. Kang, J. J. Helble, A. F. Sarofim and J. M. Beer in Proceedings of the Twenty-second International Symposium on Combustion, Pittsburgh 1988 (The Combustion Institute 1988), p. 101.
11. Gas Fluidization Technology, D. Geldart ed., Wiley, Chichester (1986).
12. J. R. Grace, Can. J. Chem. Engg. 64, 353 (1986).
13. P. Meakin and F. Family, J. Phys. A22, L225 (1989).
14. F. Family and P. Meakin, Phys. Rev. 40, 3836 (1989).
15. B. B. Mandelbrot, The Fractal Geometry of Nature, W. H. Freeman and Company, New York (1982).
16. P. Meakin, unpublished.
17. T. Vicsek and F. Family, Phys. Rev. Lett. 52, 1669 (1984).
18. R. Botet and R. Jullien, J. Phys. A17, 2517 (1984).
19. P. Meakin in Phase Transitions and Critical Phenomena, C. Domb and J. L. Lebowitz, eds., Vol. 12, p. 335, Academic Press, New York (1988).
20. M. Kolb, Phys. Rev. Lett. 62, 1699 (1989).
21. F. Family and P. Meakin, Phys. Rev. Lett. 62, 1700 (1989).
22. T. Vicsek and F. Family in "Kinetics of Aggregation and Gelation", F. Family and D. P. Landau, eds., North Holland, Amsterdam (1984), p. 111.
23. P. Meakin, T. Vicsek and F. Family, Phys. Rev. B31, 564 (1985).
24. P. Meakin and H. E. Stanley, J. Phys. A17, L173 (1984).
25. M. von Smoluchowski, Z. Phys. 17, 585 (1916).
26. M. von Smoluchowski, 92, 129 (1917).
27. Z. Cheng, S. Redner, P. Meakin and F. Family, Phys. Rev. A (1989), to be published.
28. P. Bak, C. Tang and K. Wiesenfeld, Phys. Rev. Lett. 59, 381 (1987).
29. P. Bak, C. Tang and K. Wiesenfeld, Phys. Rev. A38, 364 (1988).
30. C. Tang and P. Bak, Phys. Rev. Lett. 60, 2347 (1988).
31. C. Tang and P. Bak, J. Stat. Phys. 51, 797 (1988).
32. K. Kang and S. Redner, Phys. Rev. A30, 2833 (1984).
33. K. Kang and S. Redner, Phys. Rev. Lett. 52, 955 (1984).
34. K. Kang and S. Redner, Phys. Rev. A32, 435 (1985).
35. D. Elderfield, J. Phys. A20, L135 (1987).
36. P. D. J. van Dongen, Phys. Rev. Lett. 63, 1281 (1989).
37. P. Meakin and M. H. Ernst, Phys. Rev. Lett. 60, 2503 (1988).
38. P. Meakin, Phys. Rev. A40 2655 (1989).

Vertex Model of Cellular Pattern Growth in Two and Three Dimensions

T. Nagai[1], *K. Kawasaki*[2], *and S. Ohta*[1]

[1]Physics Department, Kyushu Kyoritsu University, Kitakyushu 807, Japan
[2]Department of Physics, Faculty of Science, Kyushu University 33, Fukuoka 812, Japan

Abstract. Growth of cellular patterns found in coarsening soap froths and grain aggregates is studied by computer simulation of a new model, the vertex model, in two and three dimensions (2D and 3D). Some results are presented and are compared with those of other studies.

1. Introduction

The problem of cellular pattern growth has been receiving attention recently in various fields [1,2]. For soap froths and grain aggregates, existing simulation models may be too detailed for simulating systems with a large number of cells. We have introduced the following model (the vertex model) which is simple enough to simulate large systems but still keeps the topological properties of cellular patterns [3,4]. We regard such systems as an assembly of vertices which are intersection points of interfaces (edges) in 2D (3D). Vertices are connected to each other through straight cell boundaries (edges) in 2D (3D). The equations of motion for vertices are obtained by balancing the friction force and the thermodynamic force. Assuming that the interface free energy is proportional to the total length (area) of cell boundaries in 2D (3D) and that the interface free energy per unit length (area) of boundary is σ and the friction constant is σ/L, we obtain the equations of motion for vertices, Eqs.(1) and (2) below. When two vertices come within a small distance two elementary processes, i.e. recombination of vertices and annihilation of cells, occur to change the topology of the pattern.

2. Two-Dimensional Vertex Model and Computer Simulation

The velocity $\mathbf{v}_i$ of the i-th vertex located at $\mathbf{r}_i$ in 2D is given by, $\mathbf{r}_{ij} \equiv \mathbf{r}_i - \mathbf{r}_j$,

$$\Sigma^{(i)}{}_j (\sigma/3L)|\mathbf{r}_{ij}|\mathbf{n}_{ij}\mathbf{n}_{ij}\cdot(\mathbf{v}_i+\mathbf{v}_j/2) = -\Sigma^{(i)}{}_j\, \mathbf{r}_{ij}/|\mathbf{r}_{ij}| \, , \qquad (1)$$

where $\mathbf{n}_{ij}$ is the unit normal vector to the straight cell boundary element <ij> and the sum is over three vertices j which are connected to i. The 2D elementary processes comprise the recombination of vertices and the annihilation of a triangular cell [3]. We have carried out computer simulations of this model and reported the results in detail in Ref.3. Here we mention it briefly.

The initial states were Voronoi cell networks with 24000 cells (48000 vertices) and the periodic boundary condition was imposed. After a short early stage, the cellular patterns generated show self-similarity (scaling), i.e. the average radius of a cell R(t) increases as $t^{1/2}$, the distribution of the cell radius R becomes a function of R/R(t) and the distribution of the number of cell edges becomes stationary at the late stage, shown in Fig.1. The edge number distributions observed in four quasi-2D real systems [5,6] are also shown in Fig.1. The agreement between the simulation and the experimental results is rather good. It has been confirmed that the data shown by the symbol Δ in Fig.1 are stationary. However, we do not know whether this is true for other experimental data cited above. Further elaborate experimental studies are required regarding this point.

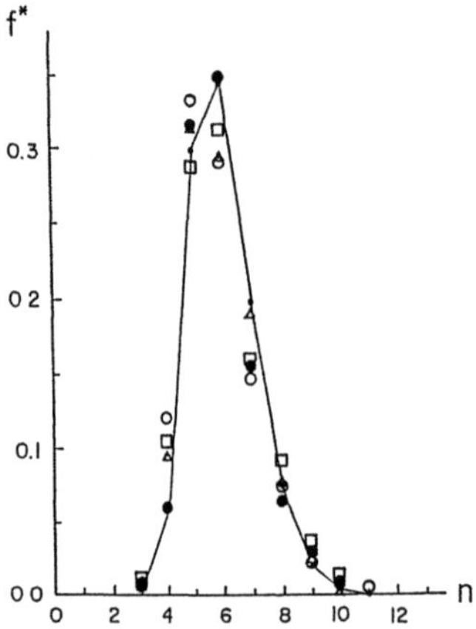

Fig.1. 2D Edge number distribution. Vertex Model:- Grains:•[5],o[5] Soap froths:Δ[6],□[5]

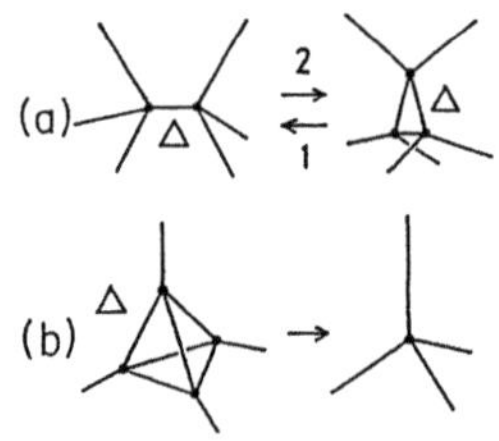

Fig.2. 3D Elementary processes.
(a)recombination
(b)annihilation

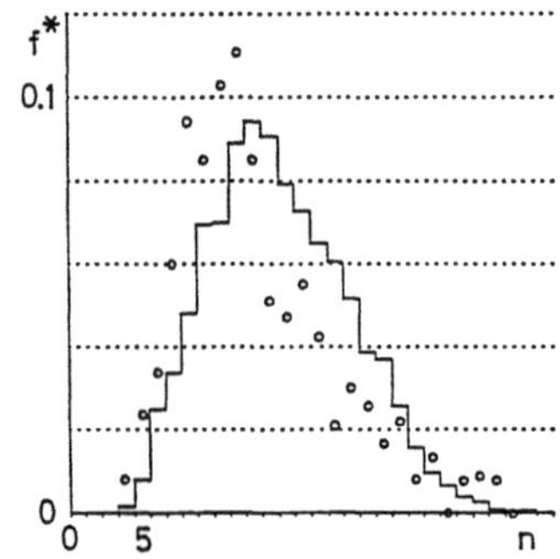

Fig.3. 3D Face number distribution. Vertex Model:- Anderson et al.[7]:o

3. Three-Dimensional Vertex Model and Computer Simulation

The velocity of the i-th vertex $\mathbf{v}_i$ in 3D is given by

$$\Sigma^{(i)}_{<jk>}(\sigma/6L)A_{ijk}\mathbf{n}_{ijk}\mathbf{n}_{ijk}\cdot[\mathbf{v}_i+(\mathbf{v}_j+\mathbf{v}_k)/2]=(1/2)\Sigma^{(i)}_{<jk>}\sigma\,\mathbf{n}_{ijk}\times\mathbf{r}_{jk}, \quad (2)$$

where j and k denote four vertices adjacent to i and the sums with respect to <jk> are over six triangles (ijk). $\mathbf{n}_{ijk}$ is the unit normal vector to the triangle (ijk) defined in such a way that the vertices i, j, k are arranged in a clockwise manner when viewed in the direction of $\mathbf{n}_{ijk}$. Hence, each term on the r.h.s. of (2) acts on the vertex i so as to reduce the area of the triangle (ijk). Here we assume that the friction force acting on a vertex arises from the triangle formed by itself and its two neighbors. The 3D elementary processes comprise the two recombination processes, i.e. switching 1 and 2 (Fig.2(a)) and the annihilation process of a tetrahedron (Fig.2(b)). We have carried out computer simulations of this model, neglecting the term $(\mathbf{v}_j+\mathbf{v}_k)/2$ on the l.h.s. of (2).

The initial states were Voronoi cell networks with 343 cells (2299 vertices) and the periodic boundary condition was imposed. The average radius of a cell R(t) increases as $t^{1/2}$ at the late stage, which is the same as that in 2D. The average number of faces of a cell decreases from 15.41 at t=0 and approaches a stationary value of 13.59 at the late stage, though it is accompanied by fluctuations due to the finite size of the system. The distribution of the number of faces of a cell tends to be stationary at the late stage. Thus we obtain the stationary distribution averaged over the scaling regime $10<t<117$ shown in Fig.3. We also give another simulation result [7] for comparison in Fig.3.

All results for the 3D vertex model strongly suggest that 3D cellular patterns show self-similar growth with a square-root growth law at the late stage. In order to confirm this we need to increase the initial number of cells and the number of runs. This is possible for the vertex model because of its great reduction in the number of degrees of freedom.

References

[1] D.Weaire and N.Rivier, Contemp.Phys.**25**,59(1984).
[2] B.Moore,C.M.Knobler,D.Broseta and F.Rondelez, J.Chem.Soc., Faraday Trans.2,**82**,1753(1986).
[3] K.Nakashima,T.Nagai and K.Kawasaki,J.Stat.Phys.**57**,759(1989).
[4] K.Kawasaki, T.Nagai and K.Nakashima, in Cooperative Dynamics in Complex Physical Systems,ed.H.Takayama(Springer-Verlag,Heidelberg,1989).
[5] C.J.Simpson,C.J.Beingessner and W.C.Winegard,Trans.Met.Soc.AIME **239**,587(1967).
[6] J.Stavans and J.A.Glazier,Phys.Rev.Lett.**62**,1318(1989).
[7] M.P.Anderson,G.S.Grest and D.J.Srolovitz,Phil.Mag.**B59**,293(1989).

Annihilation Kinetics of Particle–Antiparticle Systems with Long-Range Interaction

H. Toyoki

Department of Physics, Faculty of Education and Liberal Arts, Yamanashi University, Kofu 400, Japan

1. Introduction

The annihilation process of particles and antiparticles moving irreversibly is studied as a model for the kinetics of point-type topological defects. It describes the growth process of quenched systems with O(N) symmetry in D=N dimensions. For D=2, the interaction energy is widely known to depend logarithmically on the separations of the pair. Although high dimensional cases are less clear, the interaction energies associated with a single defect and with a pair of defects with equal but opposite charges have been calculated, which suggests that the energy increases linearly with the separation of the pair in any dimensions of $D \geq 3$ [1,2]. In the present report we consider a particle-antiparticle (PAP) system generally with power-type pairwise interaction, as a model of such systems.

The equation of motion is given by

$$\frac{d\mathbf{r}_i}{dt} = \sigma_i \sum_j \sigma_j \frac{\mathbf{r}_i - \mathbf{r}_j}{|\mathbf{r}_i - \mathbf{r}_j|^{\beta+1}}$$

where $\sigma_i = \pm 1$. A PAP pair are annihilated when they come within the core radius.

By a mean-field assumption (MFA), we have obtained the total number of particles which decreases as $N(t) \propto t^{-D/(\beta+1)}$ [3]. We will test this relation for various β's.

2. Numerical experiment

We have performed the simulation for the molecular dynamics of the above model in two dimensions. The initial distribution is given by quasi-random numbers and then the time evolution is calculated by the simple Euler method under the periodic boundary condition. Figure 1 shows the temporal change of $N(t)$ for the three cases $\beta = 0.5$, 1.0 and 2.0. The $\beta = 1$ case solely exhibits the behavior expected by MFA, while in the other cases the number decreases faster ($\beta = 0.5$) or slower ($\beta = 2$) than MFA.

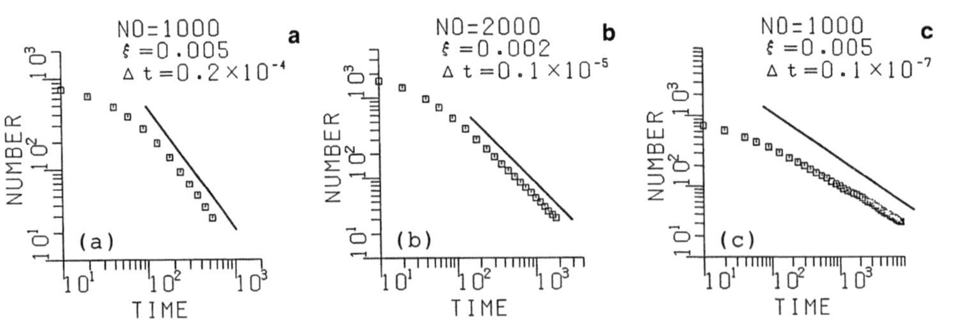

Figure 1: Number of particles vs time. (a) $\beta = 0.5$, (b) $\beta = 1.0$ and (c) $\beta = 2.0$. Time is indicated in units of Δt. The solid lines indicate the mean-field power law.

Springer Proceedings in Physics, Vol. 52 Dynamics and Patterns in Complex Fluids
Editors: A. Onuki · K. Kawasaki

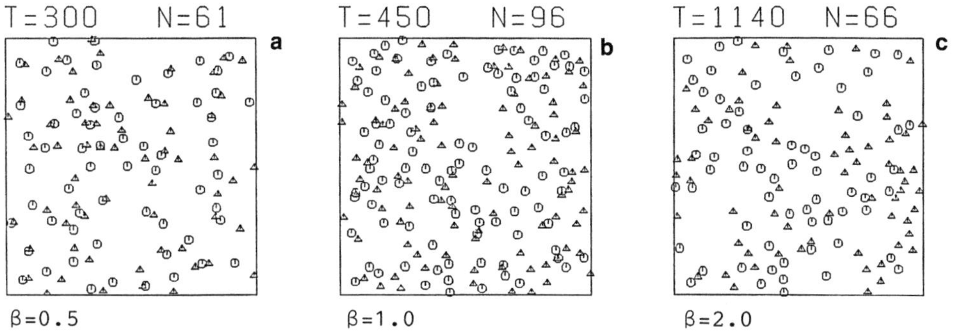

Figure 2: Particle and antiparticle positions in the intermediate stage for the three cases started from a single initial distribution.

This result can be understood qualitatively by the snapshots shown in Fig.2. These are started from a common initial distribution. In these figures, one can see that (i) the mean separation (MS) of nearest neighbor PAP pairs increases from (a) to (c) and (ii) the excess regions of particles or antiparticles appear in (c). With respect to the former phenomenon, we have found that the ratio of MS of particle-particle pairs to that of PAP remains constant only for $\beta = 1$ [2]. The latter means that the particles are annihilated locally faster than the density fluctuation diffuses away. As a result we suggest that the mean-field power law is not always satisfied, though it is satisfied in the realistic case $(D, \beta) = (2, 1)$.

3. Analytical estimation of the mean-field region

The diffusion responsible for the mean-field power law is caused by the long-range nature of the force and the randomness of the distribution. Therefore we have calculated the fluctuation of the force:

$$\sigma_f^2 = \int d^D r' d^D r'' < \delta\rho(\mathbf{r}')\delta\rho(\mathbf{r}'') > \frac{(\mathbf{r}-\mathbf{r}')\cdot(\mathbf{r}-\mathbf{r}'')}{|\,\mathbf{r}-\mathbf{r}'\,|^{\beta+1}|\,\mathbf{r}-\mathbf{r}''\,|^{\beta+1}} ,$$

where $\delta\rho$ denotes the excess density of particles. We assume that the two-point correlation of $\delta\rho$ is Gaussian: $< \delta\rho(\mathbf{r}')\delta\rho(\mathbf{r}'') >= A_0 \exp(-\,|\,\mathbf{r}'-\mathbf{r}''\,|^2 /2\ell(t)^2)$, $A_0 =< \delta\rho^2 > \propto \ell(t)^{-2D}$, and $\ell(t) \propto N(t)^{-1/D}$. As a result we have obtained the asymptotic form of σ_f:

$$\sigma_f^2 \sim \begin{cases} L^{D-2\beta}\ell(t)^{-D} & \text{for} \quad D > 2\beta & (1) \\ \ell(t)^{-2\beta} & \text{for} \quad \beta < D \le 2\beta & (2) \\ \xi^{2(D-\beta)}\ell(t)^{-2D} & \text{for} \quad D \le \beta & (3) \end{cases}$$

where L is the linear size of the system and ζ is the core radius. In the regions (1) and (3), the assumption of scaling must be invalid because σ_f depends on L or ξ. For the region (2), the diffusion length increases linearly with $\ell(t)$, which implies the validity of MFA in this region. This classification is consistent with the above numerical results.

In summary we have shown that the annihilation kinetics for (D, β) is classified into three regions and the mean-field power law is valid solely in the intermediate region. This result suggests that point defects in a quenched system might not exhibit the mean-field law for $D \geq 3$.

1. S.Ostlund, Phys.Rev. **B24**, 485 (1981).
2. H.Toyoki, submitted (1989).
3. H.Toyoki, in *Dynamics of Ordering Processes in Condensed Matter* edited by S.Komura and H.Furukawa (Plenum, 1988).

A Simple Model for the Relation Between Weight and Lifetime

S. Miyazima

Department of Physics, Chubu University, Kasugai, Aichi 487, Japan

A modified epidemiological model in which each member of the population (element) has a finite lifetime has been discussed by Bunde and Miyazima [1-3]. By introducing the lifetime of each cell, one obtains several growing clusters with characteristic growth behavior, such as those seen in the Eden model and the self-avoiding walk, depending on the lifetime τ and inhomogeneity p of the media.

Now there are very interesting data concerning the body size and lifetime of mammals, from whale to mouse, as shown in Fig.1. Generally speaking, larger animals live for longer periods. Our modified epidemics model [1-3] gives a preliminary explanation of this simple relation.

Figure 4 in [3] tells us that there are three different types of growth in the modified epidemics. For the case of long lifetime and a high density of elements, the manner of growth is equivalent to the Eden model which never stops. On the other hand, only finite clusters appear for a density of elements lower than a threshold value p_c. In the intermediate region with $p > p_c$ but a shorter lifetime of each element, we have a growth of self-avoiding-walk type, where the cluster could grow infinitely, but it actually stops growing because the advancing growth edge comes into the cavity of its own cluster. If we calculate more details in this region, we obtain the relation shown in Fig.2 between the cluster size and the lifetime of the cluster, which is very similar to Fig.1.

In the present model we have not taken into account some inefficiency which appears in larger animals. Therefore the present curve in Fig.2 may continue in a straightforward way. However, the actual data in Fig.1 show a tendency to saturate for larger animals.

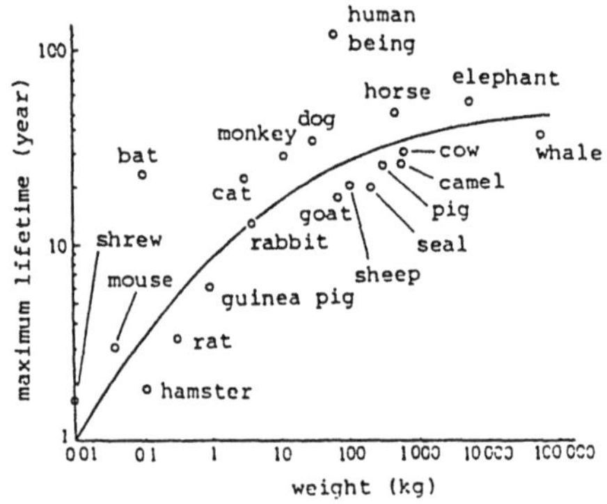

Fig.1. Relation between weight and lifetime in mammals cited from K.Suzuki "Origin of Aging" (in Japanese), Shokabo, 1988.

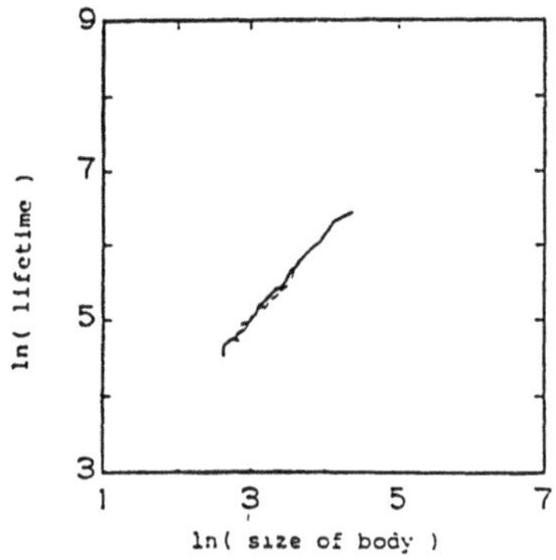

Fig.2. Relation between weight and lifetime calculated from a modified epidemiological model with p=0.8.

Springer Proceedings in Physics, Vol.52 Dynamics and Patterns in Complex Fluids
Editors: A. Onuki · K. Kawasaki

References

1. A.Bunde, S.Miyazima and H.E.Stanley, J. Stat. Phys. 47, 1(1987).
2. S.Miyazima, A.Bunde, S.Havlin and H.E.Stanley, J. Phys. A: Math. Gen. 17, L1159(1986).
3. A.Bunde and S.Miyazima, J. Phys. A: Math. Gen. 21, L345(1988).

Dynamics of Ordering Processes in a Long-Range Exchange Model

H. Hayakawa, T. Koga, and K. Kawasaki

Department of Physics, Kyushu University 33, Fukuoka 812, Japan

Abstract. A new Ising dynamics with long-range exchange is proposed in which the exchange probability is a power law function of the distance between exchanged spins. Using a coarse-grained picture, a new time-dependent Ginzburg-Landau (TDGL) equation is derived and a coarsening process is discussed.

Recently, the dynamics of ordering processes undergoing a first-order phase transition has been studied extensively. Although the exchange dynamics of the Ising model which involves exchanges between nearest-neighbor spins has been often used for a conserved order parameter, the local conservation of the order parameter is sometimes not obvious in more general situations. Therefore, a theory of a long-range spin-exchange model should be developed to describe systems in which the order parameter is globally conserved but locally nonconserved. Here, we introduce a model in which the exchange probability $p(\mathbf{r}_1, \mathbf{r}_2)$ is a power law function of the distance between exchanged spins at $\mathbf{r}_1$ and $\mathbf{r}_2$ as $p(\mathbf{r}_1, \mathbf{r}_2) \propto |\mathbf{r}_1 - \mathbf{r}_2|^{-\mu-d}$, where d is the spatial dimension and $0 < \mu < 2$[1]. Using a coarse-grained picture, it can be shown that our model belongs to the same universality class as that of a noninteger derivative model for TDGL equation[2] which is given by

$$\frac{\partial S_k(t)}{\partial t} = -L|\mathbf{k}|^{\mu} \frac{\delta F\{S\}}{\delta S_k(t)} + \zeta_k(t) \tag{1}$$

in the Fourier space, where $S_k(t)$ and $F\{S\}$ are the Fourier component of the order parameter and the free energy functional, respectively. The noise $\zeta_k(t)$ satisfies $< \zeta_k(t)\zeta_q(\tau) >= 2L|\mathbf{k}|^{\mu}\delta(\mathbf{k}+\mathbf{q})\delta(t-\tau)$.

We consider a phase separation or a coarsening process initiated by a temperature quench using the long-range exchange model(Fig.1). In the late stage of the coarsening, the local order parameter has its equilibrium values $\pm M$ except in the region of the interface. In this case, the equation of motion is determined by the motion of interfaces[3]. As was indicated by Onuki[2] the system has characteristics of a nonconserved system and the interface velocity is determined by its local curvature for $0 < \mu < 1$. On the other hand, for $1 < \mu < 2$, the equation of motion is determined by a nonlocal dynamics and has the following form for spherical droplets

Springer Proceedings in Physics, Vol. 52 Dynamics and Patterns in Complex Fluids
Editors: A. Onuki · K. Kawasaki

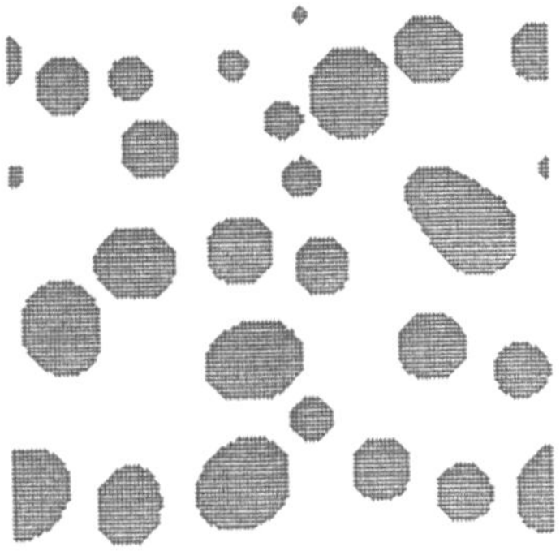

Fig.1: A snap shot of the coarsening process for $\mu = 1$ and $d = 2$ obtained by a numerical simulation.

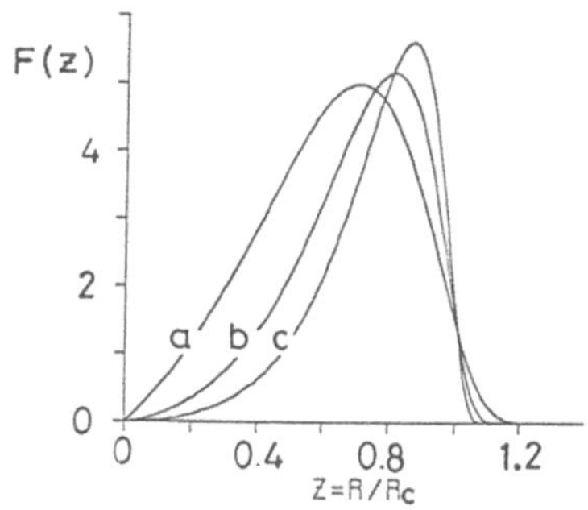

Fig.2: The μ dependence of the scaling function $F(z)$ for $\phi \to 0$ and $d = 3$, where a, b and c correspond to $\mu \to 1.0$, $\mu = 1.5$ and $\mu \to 2.0$, respectively.

$$R_i^{\mu-1}\dot{R}_i + \sum_{j \neq i} \frac{R_j^{d-1}\dot{R}_j}{|\mathbf{r}_i - \mathbf{r}_j|^{d-\mu}} = \frac{(d-1)\sigma L}{4M^2 C(d,\mu)}\left(\frac{1}{R_c} - \frac{1}{R_i}\right), \tag{2}$$

where $\mathbf{r}_i$ and R_i are the center of mass and the radius of the i-th droplet, respectively, R_c is the critical droplet radius, σ is the surface tension and $C(d, \mu)$ is a coefficient which depends on d and μ. We have ignored the effect of the noise to derive (2). Combining (2) with the equation of mass conservation and continuum equation of the size distribution, we find the distribution function $f(R, t)$ has the scaling form $f(R,t) = t^{-(d+1)/(1+\mu)} F(R/R_c)$ for $1 < \mu < 2$, where $F(z)$ is the scaling function and $R_c \propto t^{1/(1+\mu)}$. The explicit form of the scaling function for low volume fraction, ϕ, can be obtained using numerical integration. Figure 2 shows the μ dependence of the scaling function for $\phi \to 0$, where the functional form is identical to that of the interface kinetic limited system[4] for $\mu \to 1$, whereas the form coincides with that obtained by Lifshitz and Slyozov[5] for $\mu \to 2$.

[1]. H.Hayakawa and F.Family, *preprint.*
[2]. A.Onuki, *Prog.Theor.Phys.*74,1155 (1985).
[3]. K.Kawasaki and T.Ohta, *Physica* **118A**, 175(1983).
[4]. J.A.Marqusee and J.Ross, *J.Chem.Phys.*79,373(1983).
[5]. I.M.Lifshitz and V.V.Slyozov, *J.Phys.Chem.Solids.*19,35(1961).

Part VII

Pattern Formation

Dendrites, Viscous Fingers, and the Theory of Pattern Formation

J.S. Langer

Institute for Theoretical Physics, University of California,
Santa Barbara, CA 93106, USA

This lecture is based in large part on a general article with the same title published in Science magazine [1]; what follows is a summary excerpted from that article. More detailed and technical presentations of this material can be found in recent reviews [2,3,4].

The theory of pattern formation in nonlinear dissipative systems has taken some surprising turns in the last several years. One of the most interesting developments has been the discovery that weak capillary forces act as singular perturbations which lead to beautifully delicate and very nearly identical selection mechanisms both in dendritic crystal growth and in the fingering patterns which emerge when a viscous fluid is displaced by a less viscous one. It now appears likely that important progress has been made, but pieces of the puzzle still seem to be missing.

In the conventional thermodynamic model of the solidification of a pure substance from its melt, the fundamental rate-controlling mechanism is the diffusion of latent heat away from the interface between the liquid and solid phases. The latent heat that is released in the transformation warms the material in the neighborhood of the solidification front and must be removed before further solidification can take place. This is a morphologically unstable process which characteristically produces dendrites, that is, tree-like or snowflake-like structures. In a typical sequence of events, an initially featureless crystalline seed immersed in an undercooled melt develops bulges in crystallographically preferred directions. The bulges grow into needleshaped arms whose tips move outward at constant speed. These primary arms are unstable against sidebranching and the sidebranches, in turn, are unstable against further sidebranching, so that each outward growing tip leaves behind itself a complicated dendritic structure.

The hydrodynamic analog of dendritic solidification is the fingering instability that occurs when one causes an inviscid fluid to drive a viscous one through a porous medium. The two systems are formally almost identical in their theoretical descriptions because the pressure in the viscous fluid plays almost exactly the same role as the thermal field in the solidifying melt. The important two-dimensional version of the hydrodynamic situation is a Hele-Shaw cell in which the two immiscible fluids are constrained to move in a long channel of fixed width between narrowly separated parallel plates. Under these circumstances,

Springer Proceedings in Physics, Vol. 52 **Dynamics and Patterns in Complex Fluids**
Editors: A. Onuki · K. Kawasaki

an initially flat interface between the two fluids becomes unstable, and the steady-state configuration of the system is one in which a finger of the inviscid fluid moves down the center of the channel. If pushed fast enough, this finger becomes unstable against oscillatory deformations near its tip.

There are sharply defined problems of pattern selection associated with both of these systems. In solidification, it is known that the growth rate and the tip radius of a dendrite are determined uniquely by the undercooling of the melt. In the hydrodynamic case, the ratio of the width of the finger to the width of the channel is determined uniquely by the flow rate. The problem of predicting this ratio was posed originally by Saffman and Taylor. In both cases, surface tension appears at first glance to be a negligable perturbation; the lengths associated with capillarity are orders of magnitude smaller than other characteristic length scales in these systems. However, the omission of surface tension in either case leads to continuous families of solutions of the steady-state equations of motion and, thus, to no explanation whatsoever of the experimentally observed selection principles.

The central result of the recent analyses is that these continuous families of steady-state solutions do not survive in the presence of surface tension. A nonvanishing capillary constant, no matter how small, reduces the continuum of solutions to, at most, a discrete set; and the existence of any solution at all depends on there being some anisotropy in the system – a preferred set of crystallographic axes on the one hand, or interaction with nearby walls of the channel on the other.

A natural guess, then, is that the selected dendrites or fingers are the ones for which stable solutions exist. In more formal language, we guess that the condition for solvability of the steady-state equations is equivalent to a condition for the existence of a stable fixed point with a large basin of attraction in the space of configurations of the relevant dynamical system. If this conjecture is correct, orderly dendritic growth does not occur in isotropic materials. In suitably anisotropic situations, a growing body of analytic and computational evidence indicates that there is a denumerably infinite set of solutions, and that only one of these solutions – usually the fastest or the narrowest – can be dynamically stable. The hypothesis that this unique solution exists and that it describes the dynamically selected state of the system has come to be known as the "solvability theory".

The mathematical development of the solvability theory is quite technical, involving, in its deepest and most elegant version, analytic continuation of the steady-state solution to the neighborhood of its complex singularities. The central feature of the result is that the existence or nonexistence of a physically acceptable solution is governed by a term of order

$$\Theta(\sigma) \sim \exp\left(-\frac{\text{constant}}{\sqrt{\sigma}}\right)$$

where σ is a dimensionless group of parameters which is proportional to the surface tension, and which is expected to be small. The quantity $\Theta(\sigma)$ may be interpreted as a measure of the accuracy to which one must perform, say, a numerical calculation in order to determine whether a solution exists. Equivalently, $\Theta(\sigma)$ is an estimate of the strength of a perturbation which will destroy or otherwise make a qualitative change in the nature of such a solution. Because all derivatives of $\Theta(\sigma)$ are zero for vanishing positive σ, the solvability condition cannot be determined by any regular perturbation expansion in powers of σ.

Experimental tests of the solvability theory are encouraging but inconclusive at present, and they seem likely to remain in such a state for some time. The difficulty is that the cleanest experiments are those which measure growth rates of dendrites with fully three dimensional crystalline anisotropy. In my opinion, the extremely difficult theoretical computations that are required in order to make quantitative predictions for that situation have not yet been carried out satisfactorily.

Perhaps the most convincing qualitative confirmation of the theory is the bubble effect discovered by Couder and coworkers. By trapping small bubbles at the tips of Saffman-Taylor fingers, these investigators have verified predictions that certain very weak perturbations can make specific qualitative changes in the behavior of the system. For example, they have confirmed that fingers with bubbles can have widths appreciably smaller than the lower limit allowed for unperturbed fingers. Moreover, in a circularly symmetric geometry where radial fingers ordinarily suffer tip-splitting instabilities, they have shown that fingers with bubbles at their tips behave very much like dendrites, complete with sidebranches.

An as yet less complete conceptual development arising from the solvability theory is a growing understanding of the non-steady dynamics of pattern forming systems. In particular, we are beginning to understand the stability of dendritic tips and the manner in which perturbations of these tips are amplified to form complex arrays of sidebranches. One possibility that is suggested by recent results is that the sidebranches are generated by selective amplification of ambient noise – perhaps just the thermal fluctuations in the solidifying material.

A large number of both fundamental and practical questions remain to be answered. At the most basic level, there is the question of the validity of the solvability theory itself. Might this theory be mathematically correct but physically irrelevant? Perhaps real systems simply ignore the steady-state solutions and find other oscillatory, or even irregular, states of motion. Might irregular motion of dendritic tips be induced by addition of impurities at small concentrations such that thermal and solutal effects compete with one another? Might sufficiently strong crystalline anisotropy have the same effect?

At a more practical level, it will be important to understand the behavior of the arrays of dendrites that constitute the so-called "mushy zone" in solidifying, cast alloys, and which determine the metallurgical microstructures of such materials. What growth conditions control the primary spacing between these dendrites? What is the undercooling at the front of the mushy zone? Related questions have to do with cellular structures that are formed in directional solidification, and with the spacings and regularity of lamellar or rodlike eutectics. Answers to these questions, and a large number of similar ones, are needed in order to make quantitative predictions relevant to a wide variety of materials processing technologies.

References

1. J.S. Langer, Science **243**, 1150 (1989).
2. J.S. Langer, "Lectures in the Theory of Pattern Formation", in Chance and Matter, J. Souletie, J. Vannimenus, and R. Stora, eds. (North-Holland, New York, 1987) pp. 629-711.
3. D. Kessler, J. Koplik and H. Levine, Adv. Phys. **37**, 255 (1988).
4. Dynamics of Curved Fronts, P. Pelce, Ed. (Academic Press, New York, 1988).

Pattern Formation in Growth of Snow Crystals: Hexagonal and Dendritic Patterns

E. Yokoyama and T. Kuroda

Institute of Low Temperature Science, Hokkaido University, Sapporo 060, Japan

Abstract. We simulate the pattern formation in growth of snow crystals, i.e. ice crystals growing from the vapor, by taking into account (1) the surface kinetic process and (2) diffusion process under various growth conditions. Hexagonal patterns are obtained starting from a circular crystal for small supersaturation. On the other hand, dendritic patterns are obtained for large supersaturation. The conditions for the transition from hexagonal to dendritic patterns are given by the supersaturation and the dimensionless crystal size with reference to the mean free path of a water molecule in the vapor.

1. Introduction

The external form of a snow crystal changes with time under given growth conditions of temperature T and supersaturation σ, and the complicated pattern of the crystal as a whole is formed during its growth.

A spherical single ice crystal of the order of 1-10μm in radius is formed by the freezing of a supercooled water droplet in a cloud; this frozen droplet adsorbs supersaturated water vapor, thus growing into a hexagonal prism, which is the fundamental form of ice crystals, bounded by two basal and six prism faces. At low supersaturation the hexagonal prism can grow in a stable way, retaining its form. With increasing supersaturation, however, the hexagonal prism develops into a dendrite at a temperature range between $-10°$ and $-22°$C or into a needle at a temperature range between $-4°$ and $-10°$C by preferred growth of edges and corners of the crystal.[1]

In this study, we investigate pattern formation in the growth of snow crystals under various growth conditions of supersaturation and diffusion coefficient of water vapor by taking into account the following two processes relevant to the growth of crystals: (1) a surface kinetic process for the incorporation of water molecules into the ice crystal lattice, which includes their adsorption onto a crystal surface, surface diffusion of admolecules toward the steps and the lateral motion of steps, and (2) a diffusion process for supplying water molecules in air toward the crystal surface. We analyze the pattern produced by the simulation from the standpoint of not only the growth conditions but also the crystal's size. Lastly, we discuss the mechanism of pattern formation of snow crystals in terms of processes (1) and (2).

Since this moving boundary problem in which two processes are coupled is extremely difficult to solve, we consider the time development of a two-dimensional single crystal which is initially a circle perpendicular to the c-axis of a hexagonal snow crystal.

2. Calculation Methods[2,3]

The supersaturation σ surrounding a crystal is governed by the quasi-static equation

$$\Delta\sigma=0 \ , \qquad (1)$$

subject to the following boundary conditions: (i) $\sigma = \sigma_\infty$ on the boundary, which is a circle

Springer Proceedings in Physics, Vol. 52 Dynamics and Patterns in Complex Fluids
Editors: A. Onuki · K. Kawasaki

of radius R centered on the center of the crystal, and (ii) the growth rate V_k determined by the surface kinetic process should be equal to the growth rate V_d determined by the diffusion process, i.e. $V_k = V_d$ [4]. It should be noticed that these boundary conditions are very important for determining self-consistently the growth rate V controlled both by the surface kinetic process and by the diffusion process. We solve Eq.(1) by means of the boundary element method, get V at each position on the surface at a certain moment and then determine the shape of a growing surface after Δt seconds.

3. Simulation Results and Discussions[3]

The temperature of the whole system is assumed to be uniform at $-15°$C, which is a typical temperature for the development of a dendritic pattern. The diffusion coefficient is D=0.2cm^2/s (corresponding to 1 atmosphere) and the radius of the initial crystal 5×10^{-3}cm.

The following patterns are obtained by simulation (Fig.1): (a) Hexagonal pattern for small supersaturation $\sigma_\infty = 3.5\%$: An initial circular crystal develops into a perfect hexagon (3000s), and then the hexagonal crystal can grow retaining its flat surface macroscopically. (b) Dendritic pattern for large supersaturation $\sigma_\infty = 32\%$: An initial circular crystal develops into a perfect hexagon (800s), and then the hexagon develops into a dendritic pattern by way of preferred growth of six corners.

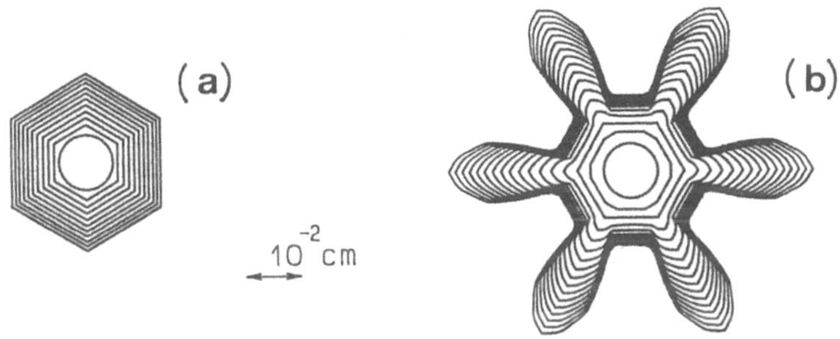

Fig. 1 (a) Hexagonal pattern for $\sigma_\infty = 3.5\%$. (b)Dendritic pattern for $\sigma_\infty = 32\%$.

The conditions for transition from the hexagonal to a dendritic pattern have been given by the supersaturation σ_∞ and the dimensionless crystal size $\mathscr{L}$ with reference to the ratio of the diffusion coefficient D to the mean speed $\bar{v}$ of water molecules. The ratio D/$\bar{v}$ is the mean free path $\bar{l}$ of a water molecule in vapor. The relation between simulated patterns and growth conditions of σ_∞ and $\mathscr{L}$ has been shown in a phase diagram of $(\sigma_\infty, \mathscr{L})$[3].

References

[1] T. Kobayashi, Phil. Mag. 6,1363(1961).
[2] E. Yokoyama and T. Kuroda, Dynamics of Ordering Processes in Condensed Matter, ed. S. Komura and H. Furukawa, (Plenum Press, New York 1988), p.95.
[3] E. Yokoyama and T. Kuroda, Phys. Rev. A 41,2038(1990).
[4] T. Kuroda, T. Irisawa and A. Ookawa, J. Cryst. Growth. 42, 41(1977).

Periodic Changes in the Structure of the Growing Crystal Surface

T. Irisawa[1], *Y. Arima*[1], *and T. Kuroda*[2]

[1]Department of Physics, Gakushuin University, Tokyo 171, Japan
[2]Institute of Low Temperature Science, Hokkaido University, Sapporo 060, Japan

Abstract. Periodic changes in the structure of a growing crystal surface are investigated theoretically by means of a Monte-Carlo simulation. The conditions for occurrence of the periodic changes in the structure of the growing crystal surface are obtained for singular surfaces.

1. Introduction

It has been recently reported that RHEED intensity for the growing surface of crystals in the process of MBE oscillates with a period corresponding to the completion of a monolayer [1].

In this study, the periodic change of the structure and the flatness of the growing surface are investigated theoretically. The growth of the (001) face of simple cubic crystal is simulated by using Gilmer and Bennema's model for vapor growth [2].

2. Two-dimensional nucleation in the case of no evaporation of adatoms

In normal vapor growth treated by the BCF theory [3], the life time τ_s of the adatom before evaporation from the surface is an important parameter.

On the other hand, the life time τ_c of an adatom before capture by another adatom is the important parameter which controls the growth mode in MBE growth, since τ_s is larger than τ_c because of the lower substrate temperature.

As an adatom migrates around in the area $D_s(cm^2/s)$ for a second, it encounters D_sN_1 adatoms per second, where N_1 is the adatom density and D_s is the surface diffusion coefficient. Therefore, τ_c is given by $\tau_c = 1/D_sN_1$. If the incident beam flux J is very large, the dissociation of dimer does not occur. Moreover, if $\tau_c<\tau_s$, the decrease in the adatom density due to evaporation from the surface can also be ignored. Therefore, the adatom density under steady state conditions is given by $N_1 = (J/D_s)^{1/2}$ [4]. Then, τ_c is expressed as

$$\tau_c = (JD_s)^{-1/2}. \quad (1)$$

The stable clusters (i>2) are created at $t \simeq \tau_c$, and are as many as $N_s = (N_1/\tau_c)\tau_c = N_1$. Therefore, the territorial area $\lambda_c^2 = 1/N_s$ of each stable nucleus is expressed as follows:

$$\lambda_c^2 = (D_s/J)^{1/2}. \quad (2)$$

3. The periodic change of the surface structure and results

When $N_s=N_1$, there is only one adatom in the territory on average, and no further nucleation by encounters of adatoms occurs. Therefore the surface structure changes periodically as follow:
a dimer is created on each territory of area λ_c^2 and the surface is completed when the dimer grows to the maximum diameter λ_c (fig. 1).

Springer Proceedings in Physics, Vol. 52 Dynamics and Patterns in Complex Fluids
Editors: A. Onuki · K. Kawasaki

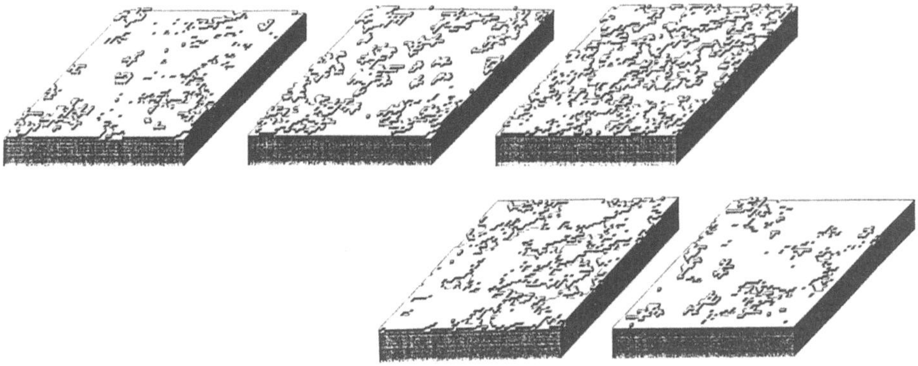

Fig. 1. Change of the surface structure from the 4th layer to 5th layer on a 64×64 substrate at $\phi/kT=20$, $\Delta\mu/kT=38$ and $\varepsilon=0.5$. Initially, the surface was flat.

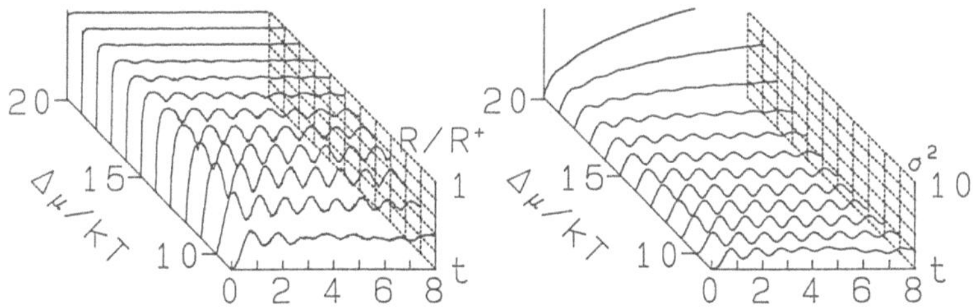

Fig. 2. Changes of (a) the normalized growth rate R/R+ and (b) the dispersion σ^2 of the height for various values of $\Delta\mu/kT$ at $\phi/kT=10$ and $\varepsilon=0.5$.

However, when incident flux J or chemical potential $\Delta\mu$ of vapor is so large that $\lambda_c \leq a$, the flatness of the surface is lost because the adatoms quickly crystallize without surface diffusion ($\Delta\mu/kT>19$ in fig. 2). From this condition, the following inequality is obtained for the (001) face of the simple cubic lattice

$$\Delta\mu \leq (3 - \varepsilon)\phi, \quad (3)$$

where ϕ is bond energy and $\varepsilon\phi=E_{sd}(0\leq\varepsilon\leq1)$ is the activation energy of surface diffusion.

On the other hand, the condition $i^*\leq1$ must be satisfied for the picture used in this study to be realized. Using a broken bond model for step energy, the following relation is derived from $i^*\leq1$:

$$\Delta\mu \geq [2\sqrt{\pi} - (1 + \varepsilon)]\phi. \quad (4)$$

Therefore, when the conditions (3) and (4) are satisfied, the surface structure of the growing crystal changes periodically, retaining flatness with no evaporation of adatoms ($19>\Delta\mu/kT>15$ in fig 2).

On the other hand, the evaporation of adatoms occurs when $\tau_c/\tau_s>>1$ for higher substrate temperature or smaller incident flux. In this case, the growth rate R itself oscillates with every completion of a monolayer because the evaporation flux of adatoms into the vapor changes with the periodic change of the surface structure ($\Delta\mu/kT<15$ in fig. 2).

References.

[1] J. J. Harris and B. A. Joyce, Surface Sci. 108 (1981) L90.
[2] G. H. Gilmer and P. Bennema, J. Appl. Phys. 43 (1972) 1347.
[3] W. K. Burton, N. Cabrera and F. C. Frank, Phil. Trans. Roy. Soc. London A243 (1951) 299.
[4] T. Irisawa, Y. Arima and T. Kuroda, Proc. ICCG-9, J. Crystal Growth (1989) in press.

Crystal Growth, Fractal Growth and Liquid-like Behavior in the Case of Thin Film Formation

S. Ozawa[1], *Y. Sasajima*[1], *and T. Haseda*[2]

[1]Ibaraki University, Faculty of Engineering, 4-12-1, Nakanarusawa, Hitachi 316, Japan
[2]Toin University of Yokohama, Faculty of Engineering, 1614, Kurogane, Midoriku, Yokohama 227, Japan

1. Introduction

The growth processes of thin films have been studied extensively not only because they are important in technology, but also because it is scientifically important to understand the condensation kinetics from the atomistic point of view. Various kinds and degrees of atomic order have been observed in the experiments of vacuum-deposited thin films. These include the long range order with/without the translational symmetry in the atomic arrangements (crystal / quasicrystal), and the short range order (amorphous thin film). The aim of this paper is to study why these different types of atomic arrangements occur in the condensation processes.

2. Monte Carlo Simulation of Growth Kinetics of Thin Films

The film formation processes including the diffusion of atoms, the nucleation, and the coalescence of nuclei are simulated by the Monte Carlo method[1,2]. We use 128×50 two-dimensional square lattice with the periodic boundary condition in the horizontal direction. We arrange 128 substrate atoms horizontally in the bottom of the lattice. We deposit a film atom onto the substrate every τ -Monte Carlo steps. And we move the deposited atoms in proportion to the Boltzmann's factor, $\exp(-\Delta H/kT)$, where ΔH is the change in the Hamiltonian associated with the transition, k and T are the Boltzmann's constant and the temperature, respectively. Our Hamiltonian is given as the sum of the values of the Morse potential of the constituent atom pairs. The dependence of the growth mode of thin films on the energy parameters has been discussed in Ref.[1].

In the present study, we carry out the Monte Carlo experiments at much larger deposition rates than those in Ref.[1] (by a factor of $10^1\sim 10^4$). Figures 1 (a)-(c) show the typical results of the dependence of the film growth on the deposition rate, $1/\tau$. The parameters used in the Morse function are $E_{fs}=0.2$eV, $E_{ff}=0.4$eV, $r_0=3.0$A, $\alpha=6$ (see Ref.[1]), and T=300K. It should be noted that the mobility of film atoms are the same in the cases (a)-(c). For reference, the result of the computer experiment in which the deposited atoms are not mobile is shown in Fig. 1(d). This corresponds to the deposition at T=0K. It is seen from these figures that the film structure is determined by two competing kinetics i.e. the liquid-like coagulation kinetics and the branch forming kinetics (the fractal pattern forming kinetics).

Springer Proceedings in Physics, Vol. 52 Dynamics and Patterns in Complex Fluids
Editors: A. Onuki · K. Kawasaki

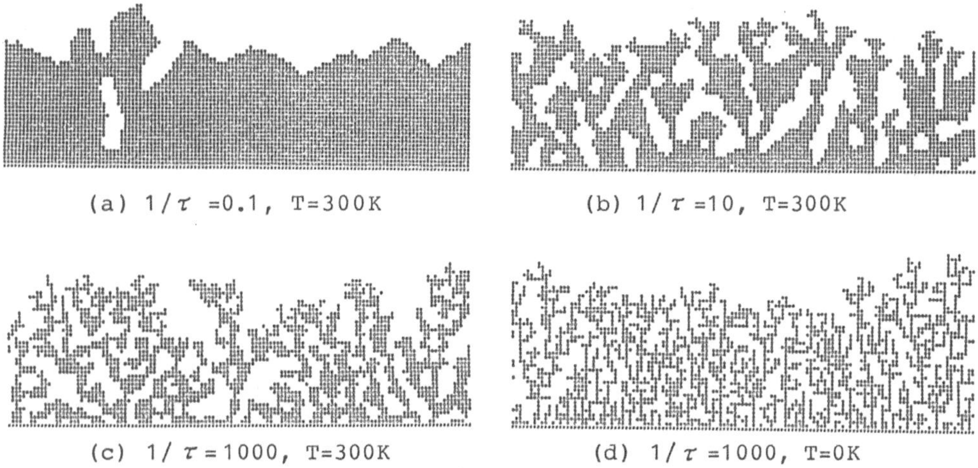

Fig. 1 Condensation of film atoms onto substrate at different deposition rates, $1/\tau$.

3. Growth of Quasicrystals

We carried out computer experiments in order to obtain Henley-Elser type quasiperiodic structure [3,4]. Figure 2(a) shows the pattern of coagulation of three kinds of atoms with different sizes and different pair binding energies; the values of these parameters are chosen so that the cluster with five-fold symmetry is most stable [3]. The atoms are sequentially added to a seed cluster shown by the filled circles in the figure in proportion to the Boltzmann's factor by mono-atomic processes. It has been found that the pattern has an amorphous structure. However, in the case where we take into account simultaneous movement of atoms of a five-fold symmetric cluster, we can obtain atomic arrangement with almost complete long range order of quasicrystals (see Fig. 2(b)).

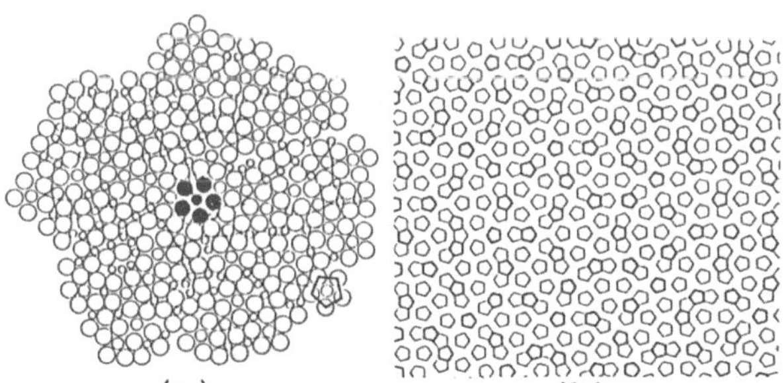

Fig. 2 (a) : Coagulation of three kinds of atoms by mono-atomic process, (b) : Coagulation of clusters, the pentagons represent the five-fold symmetric clusters. Simultaneous movement of atoms is essential for the development of quasicrystals.

References

1. S. Ozawa and Y. Sasajima: to appear in Vacuum.
2. D.W. Heermann: Computer Simulation Methods in Theoretical Physics Springer-Verlag, Heidelberg (1986).
3. Y. Sasajima, H. Sugiyama and S. Ozawa: to be submitted to FORMA.
4. C.L. Henley : Phys. Rev. B34(1986)797.

Elongation Process of Whisker Crystals with a Solution Layer

K. Kishi, T. Yoshida, and K. Yoshioka

Department of Applied Physics, Faculty of Science,
Science University of Tokyo, 1-3 Kagurazaka, Shinjuku-ku, Tokyo 162, Japan

Abstract. A solution-growth process of a whisker crystal is simulated. The crystal has a fit solution layer on its surface and grows with the vaporization of solvent. Both the atmosphere and the layer are treated as diffusion fields. It is shown that dislocation-free whisker crystals can grow in such a composite diffusion field without morphological change.

1. Introduction

According to the theories of morphological instability, needle crystals growing in a diffusion field change into dendritic shape [1]. However, many dislocation-free whiskers, which are a kind of needle crystals, grow without any branchings. One of the authors reported that the presence of a fit diffusion layer placed on a crystal surface keeps the shape of the crystal stable [2].

This paper describes a growth model of whisker crystals in a composite diffusion field where a solution layer acts as a buffer.

2. Growth Model and Computation

It is assumed that a whisker crystal standing on a heterogeneous substrate possesses a simple cubic structure bounded by {100} perfect surfaces. A saturated solution layer covers the crystal and the substrate. The shape of the solution layer on the crystal is a column with a hemispheric top and its width is kept constant.

A kind of finite element method was employed for the computation. The growth process for the computation is described as follows: At the surface of the solution layer, the solvent vaporizes according to the Gibbs-Thomson formula and therefore an excess of the solute is formed. The vapor and the excess diffuse in the atmosphere and the solution layer respectively. The excess incorporates at the surface of the crystal. Layer-by-layer growth is considered. Consumed solvent and solute are compensated from the saturated solution layer spread over the substrate. The surface diffusion process of adatoms has no effect on the growth rate since detached adatoms still remain in the solution layer.

Springer Proceedings in Physics, Vol. 52 Dynamics and Patterns in Complex Fluids
Editors: A. Onuki · K. Kawasaki

3. Results and Discussion

It is shown that the whisker crystal could grow with the singular surfaces smoothly when the vaporization rate was kept small. Results shown here are obtained under the conditions that the solubility was 2 vol.% and the critical supersaturation for two-dimensional nucleation σ_c was 10%.

In the early stage, the whisker crystal grew axially at the tip. The growth rate increased with time. Simultaneously, in the solution layer on the lateral surfaces, the degree of supersaturation increased until a nucleus formed. The nucleation caused the local decrease in the degree of supersaturation. This intermitted the axial growth, however, the decrease propagated in the solution with the advance of growth step (Fig. 1) and the axial growth restarted. The whisker crystals grew with increasing aspect ratio (Fig. 2). These results agree with the growth of actual whisker crystals.

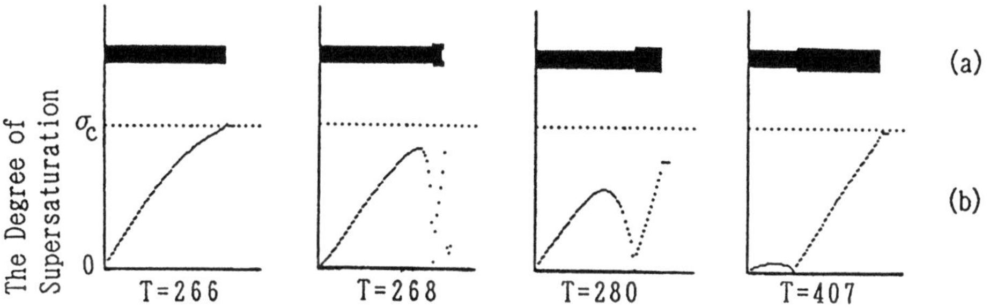

Fig. 1 The motion of growth step on the lateral surface. A growth step generated at T=266. T:Time (arb. units). (a) The shape of a whisker crystal. (b) The degree of supersaturation in the solution layer.

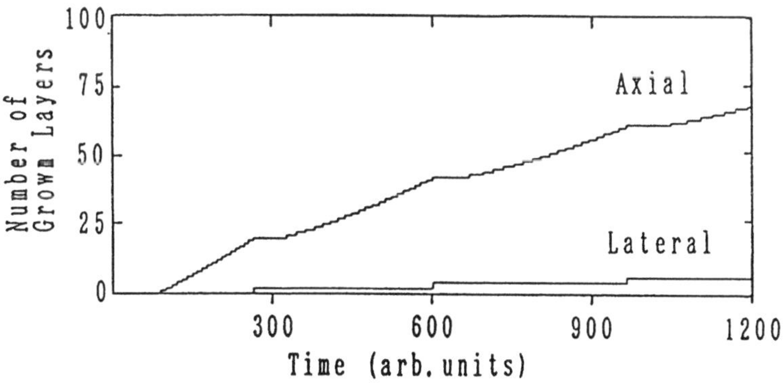

Fig. 2 Time dependence of the thickness of axial and lateral growth layers.

References

1. J. S. Langer, Rev. Mod. Phys. 52, 1 (1980) and references therein.
2. K. Kishi, A. Yoshida and K. Koike, Appl. Surf. Sci. 33/34, 525 (1988).

Anisotropic Fractal Aggregation in a Finite Density Gas

M. Uwaha[1], *Y. Saito*[2], *and S. Seki*[1]

[1]Institute for Materials Research, Tohoku University, 2-1-1 Katahira, Sendai 980, Japan
[2]Department of Physics, Keio University, 3-14-1 Hiyoshi, Kohoku-ku, Yokohama 223, Japan

Abstract. The structure and dynamics of an aggregation from a finite density lattice gas are studied. A characteristic length scale ξ, which is determined by the fractal dimension of the diffusion-limited aggregation, governs the structure of the solid, the growth velocity and the interface width. The introduction of the lattice anisotropy changes the scaling behavior.

The diffusion-limited aggregation(DLA)[1] is the low density limit of growth in a diffusing gas, which is often seen in the real world, but its relevance to the dynamics of diffusion growth is still unclear. By Monte Carlo simulation we study the structure and dynamics of an aggregation when the aggregate grows from a linear seed in a lattice gas of nonzero density n_g[2]. Our model naturally interpolates between the DLA and the Eden model. Because of the one-dimensional geometry of the system, a steady growth is realized (Fig.1a), which allows us to make simple scaling arguments.

At a low n_g and for a short length scale up to a characteristic length ξ, the structure of the aggregate looks like a fractal similar to the DLA. For a large length scale it is compact and has a nonzero asymptotic density[3]. The crossover length ξ from fractal to compact structure satisfies the relation $\xi^{D_f-d} \sim n_g$, which states that the average density of a DLA within a range of ξ is the same as the gas density n_g. Here D_f is the fractal dimension of DLA in a d-dimensional space. Note the analogy to the critical phenomena where the density n_g plays the role of temperature.

The steady growth rate V is inversely proportional to the diffusion length. If we assume that the diffusion length in the gas is the same as the characteristic length ξ in the solid, the growth velocity is given by $V \sim \xi^{-1} \sim n_g^{1/(d-D_f)}$. This is confirmed in the simulation (the curve m=1 in Fig.1b) with D_f=1.71 for d=2.

The scaling behavior of the width w of the growing region is also investigated. The growing interface can be considered as a randomly fluctuating chain of blobs with the size ξ. This random walk picture predicts $w \sim \xi(L/\xi)^{1/2} \sim n_g^{-1/2(d-D_f)} L^{1/2}$, for a system with width L, in consistency with our simulation result[2].

As for the structure of the solid, we made a multifractal analysis by plotting $(q-1)^{-1}\ln(\sum_i p_i^q)$ vs $\ln b$, where $p_i = \sum_{l\in i} n_l/N$ is the number of atoms in an i-th box with a linear dimension b divided by the total number of atoms N in the analyzed region (We analyze only the uniform region to avoid the edge effects). The slopes D_q of the graphs are 1.65 $\sim$ 1.57 for q = 0 $\sim$ 32 which is the same as the DLA type simulation[4] but smaller than the expected value 1.71. The small value may be due to an autocorrelation. Thus we analyze correlation integrals instead of the

Springer Proceedings in Physics, Vol. 52 Dynamics and Patterns in Complex Fluids
Editors: A. Onuki · K. Kawasaki

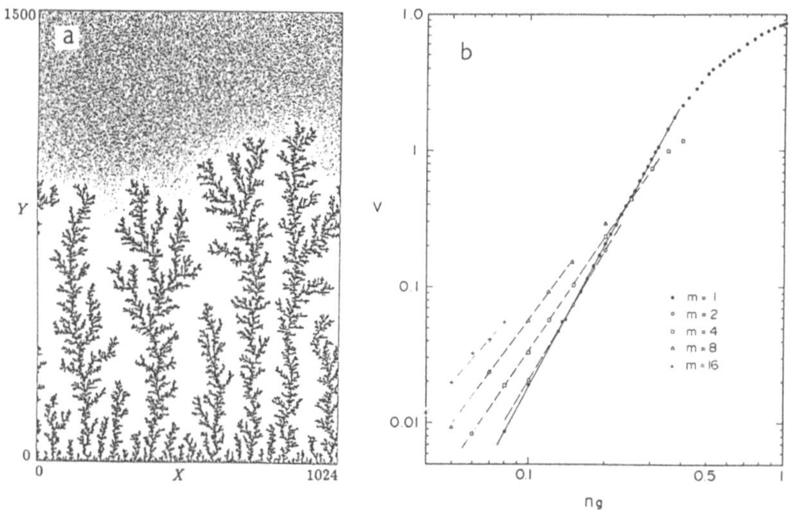

Fig.1 a) Configuration of an aggregation and surrounding gas atoms.
b)Growth rate V versus gas density n_g for different values of m.

moments of p_i to make this effect smaller. For example, for $q = 2$, $\sum\sum_{l\neq m,\in i} n_l n_m = (\sum n_l)^2 - \sum n_l = N^2 p_i^2 - N p_i$. The subtraction of the second term eliminates the autocorrelation. The resultant values of $D_{2,4}$ are closer to 1.7. Hence we conclude that the structure of the aggregate is a simple fractal similar to that of DLA.

By allowing solidification only on a mesh of the size m[5], the effect of anisotropy is emphasized, which is relevant to the dendritic crystal growth. The aggregation pattern becomes like a dendritic crystal as the anisotropy increases[6]. Simultaneously, the effective fractal dimension obtained by the V-n_g relation decreases as is shown in Fig.1b. These features are in common with the DLA models[7] as expected.

We also analyzed the density correlations in the x and y directions in the solid. They both show power law decay in short distances. However, we find a slight difference in the powers for $m = 1$ (original model): The decay is slower in the y-direction. As m increases this tendency becomes more prominent. Unfortunately we cannot obtain accurate quantitative data because of the limited computational ability.

References

[1] T.A. Witten and L.M. Sander, Phys. Rev. Lett. **47**, 1400 (1981); Phys. Rev. B **27**, 5686 (1983).
[2] M. Uwaha and Y. Saito, J. Phys. Soc. Jpn. **57**, 3285 (1988): Phys. Rev. A **40**, in press.
[3] R.F. Voss, J. Stat. Phys. **36**, 861 (1984).
[4] F. Argoul, A. Arneodo, G. Grasseau and H.L. Swinney, Phys. Rev. Lett. **61**, 2558 (1988).
[5] M. Matsushita and H. Kondo, J. Phys. Soc. Jpn. **55**, 2483 (1986).
[6] M. Uwaha and Y. Saito, J. Cryst. Growth (in press).
[7] J. Kertesz and T. Vicsek, J. Phys. A **19**, L257 (1986), J. Nittmann and H.E. Stanley, Nature **321**, 663 (1986).

Multifractal Structure of Non-Newtonian Viscous Fingers

T. Nagatani[1] *and Y. Usami*[2]

[1]College of Engineering, Shizuoka University, Hamamatsu 432, Japan
[2]Institute of Physics, Kanagawa University, Rokkakubashi, Yokohama 221, Japan

A real-space renormalization-group method has been developed to study the fractal and multifractal structures in diffusion limited aggregation [1-4]. In this article, we investigate the multifractal structure of the growth probability distribution in the viscous fingering of a non-Newtonian displaced fluid. The dependences of the α-f spectra of the growth probability distribution are shown on the parameter k describing the different non-Newtonian fluids.

The velocity and pressure fields are given by

$$v=-K\,|\mathrm{grad}\ P|^{k-1}\mathrm{grad}\ P, \quad (1)$$

$$\nabla(|\nabla P|^{k-1}\nabla P)=0 \text{ for the non-Newtonian displaced fluid,} \quad (2)$$

where v is the velocity, P the pressure,and K the permeability coefficient. The parameter k represents the strength of non-Newtonian property. For Newtonian fluids, we have k=1. Because of the $(\mathrm{grad}\ P)^k$ term in (1), the preferential growth of the tips is more pronounced in the non-Newtonian case k>1 than for the Newtonian case.

We use an electrostatic analogy to transform the viscous fingering problem into a specific type of resistor network problem. The material has a characteristic property of a nonlinear resistor. The current density j and the electrostatic potential Φ satisfy

$$j=-\sigma\,|\mathrm{grad}\ \Phi|^{k-1}\mathrm{grad}\ \Phi, \quad (3)$$

$$\nabla(|\nabla\Phi|^{k-1}\nabla\Phi)=0. \quad (4)$$

Due to the formal similarity between eqs.(2) and (4), the fluid flow problem becomes isomorphic to the electric problem.

We consider the dielectric breakdown model on the nonlinear-resistor network. For simplicity, we consider the diamond hierarchical lattice on which each bond consists of a nonlinear resistor. The nonlinear resistor has the characteristic current-voltage relation

$$I=\sigma\,|V|^k\mathrm{sgn}\ V. \quad (5)$$

A constant voltage is applied between the bottom and the top on the diamond hierarchical lattice. Then the nonlinear resistor network problem is solved. A growth probability proportional to the current is assigned to the perimeter bond.

We apply the renormalization-group method to the system. All the space of the diamond lattice is partitioned into cells of size b=2,each containing a single gerator. After a renormalization transformation these cells play the role of renormalized bonds. We assign an effective conductance $\sigma_{0,n}\sigma_s$ for the growth bond on the perimeter. Then the conductance of the cell is renormalized to $\sigma_{0,n-1}\sigma_s'$. Here $\sigma_{0,n}$ is the conductance of the unbroken bond on the nth generation. We call the conductance σ_s the surface conductance. When k=1, the surface con-

Springer Proceedings in Physics, Vol. 52 Dynamics and Patterns in Complex Fluids
Editors: A. Onuki · K. Kawasaki

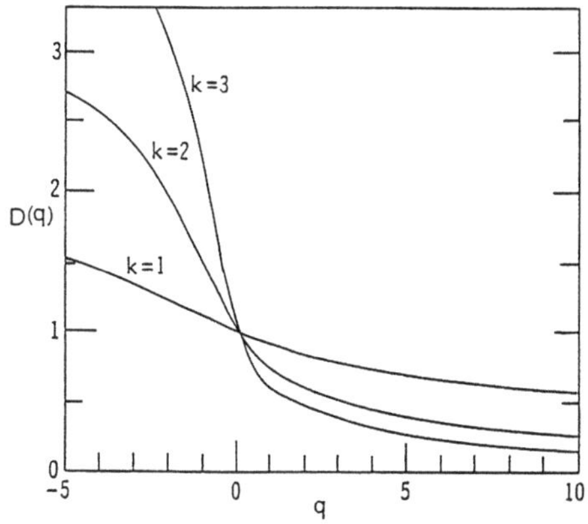

Fig.1 The generalized dimension D(q) of the growth probability distributions. The curves indicate, respectively, the results for k=1,2 and 3.

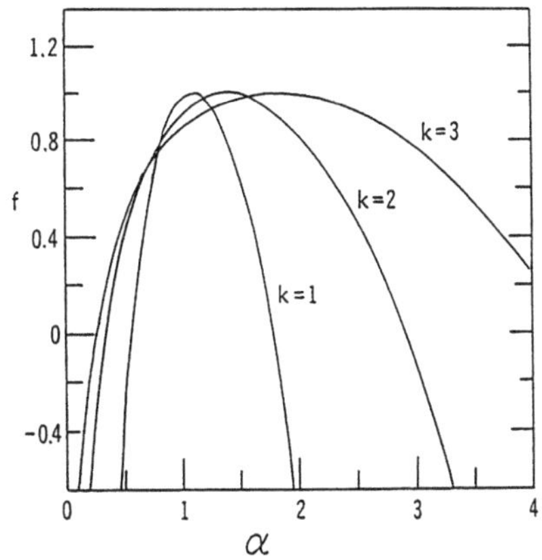

Fig.2 The α-f spectra of the growth probability distribution obtained from D(q) in Fig.1.

ductance is consistent with that of the linear resistor network problem. We obtain the renormalization function: $\sigma_S' = R(\sigma_S)$. We find a stable fixed point σ^* from $\sigma^* = R(\sigma^*)$ for a k value. An infinite hierarchy of the generalized dimension D(q) is given by

$$D(q) = -(q-1)^{-1} \ln\left[\sum_\alpha C_\alpha^* \sum_i (p_{\alpha,i}^*)^q\right] / \ln b, \tag{6}$$

where C_α^* is the configurational probability evaluated at the fixed point, and $p_{\alpha.i}^*$ the growth probability of the cell evaluated at the fixed point. The values of D(q) are plotted in Fig.1 for k=1,2 and 3. The α-f spectra of the growth probability distribution obtained from D(q) are displayed in Fig.2. It is found that the non-Newtonian fluid has important effects on the multifractal structure of the growth probability distribution.

References

1. T.Nagatani, J.Phys.A20, L381(1987); Phys.Rev.A36, 5812(1987); Phys.Rev.A38, 2632(1988).
2. L.Pietronero, A.Erzan and C.Evertz, Phys.Rev.Lett.61, 861(1988); Physica A151, 207(1988)
3. X.R.Wang, Y.Shapir and M.Rubinstein, Phys.Rev.A39, 5974(1989); J.Phys.A22, L507(1989)
4. J.Lee, P.Alstrom and H E Stanley, Phys.Rev.A39, 6545(1989)

A Crossover Phenomenon in Electrochemical Deposition

Y. Usami[1] *and T. Nagatani*[2]

[1]Department of Applied Physcis, Tokyo Institute of Technology, Meguroku, Tokyo 152, Japan
[2]College of Engineering, Shizuoka University, Hamamatu 432, Japan

A crossover phenomenon from DLA fractal to dense aggregation in electrochemical deposition is studied by means of computer simulation. The model equations are expressed in terms of the coupled diffusion fields [1]; one is the concentration field of ions C($\mathbf{x}$), and the other is the electric potential field $\Phi(\mathbf{x})$:

$$\begin{cases} \nabla \cdot (\alpha C(\mathbf{x}) \nabla \Phi(\mathbf{x}) + \nabla C(\mathbf{x})) = 0, \\ \nabla^2 \Phi(\mathbf{x}) = 0, \end{cases}$$

where α represents the strength of applied voltage. Each field is normalized to be zero on the surface and unity at the outer boundary. The algorithm of the growth process is similar to that of the DLA presented by Niemeyer *et al.* [2].

Fig.1 shows the patterns generated by this model. The crossover from DLA fractal to the dense aggregate is found on increasing the α, i.e. the applied voltage. These results are similar to experimental ones [3]. The mechanism of the aggregation presented here is more realistic when compared with experiments on the electrochemical deposition than a drift-DLA model [4].

Fig.2 shows the contour map of the concentration field as a colored graph. With increasing voltage, we can observe that the colored area becomes small around the aggregate and the area of low concentration, colored yellow, purple, and red, decreases. This reflects the distortion of the concentration field towards the aggregate and even towards the inside of fjords of the aggregate.

Fig.3 shows the magnitude of the growth probability distribution at perimeter sites. We can see that perimeter sites inside fjords come to have a large growth probability with increasing voltage. This is the reason for the change of the morphology of the aggregate from DLA fractal to the dense aggregate.

Numerical calculations were performed by HITAC-S820 at the Institute for Molecular Science's computer center.

References

1. Y.Usami and T.Nagatani: J.Phys.Soc.Jpn.**59**,474(1990).
2. L.Niemeyer, L.Pietronero, and H.J.Wiesmann: Phys.Rev.Lett.**52**,1033(1984).
3. D.B.Hibbert and J.R.Melrose: Phys.Rev.**A38**,1036(1988)
4. P.Meakin: Phys.Rev.**B28**,5221(1983).

Springer Proceedings in Physics, Vol. 52 **Dynamics and Patterns in Complex Fluids**
Editors: A. Onuki · K. Kawasaki

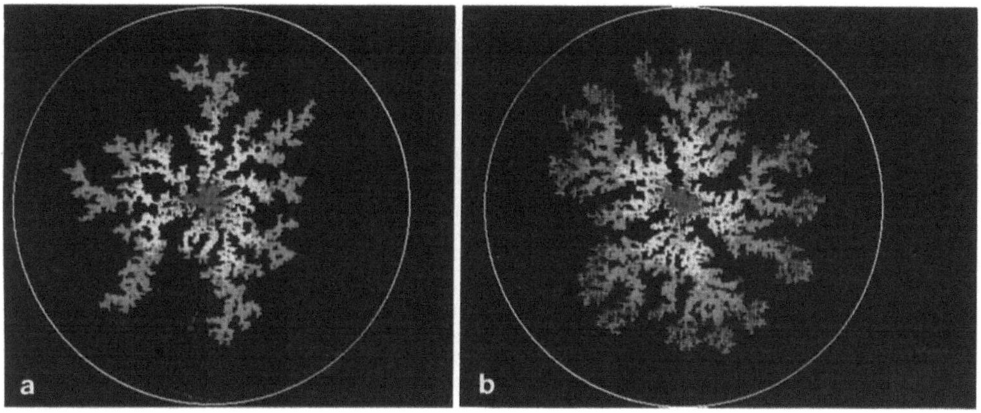

Fig.1 Typical clusters grown by the coupled diffusion fields for the parameters (a) α=0, (b) α=35. The colors indicate order of attachment to the aggregate. The system size is 400 lattice units.

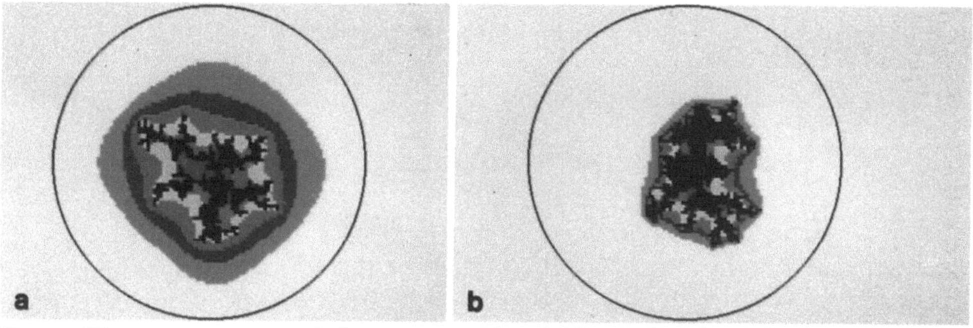

Fig.2 The contour maps of the concentration field as colored graphs for the parameters (a) α=0, (b) α=10. The colors represent the magnitude of the concentration field C(x): $0 \sim 1.0 \times 10^{-4}$, red; $\sim$0.005, purple; $\sim$0.05, yellow; $\sim$0.25, green; $\sim$0.5, blue; and $\sim$0.8, sky blue, respectively. The black region located inside the colored region represents the aggregate. The system size is 100 lattice units.

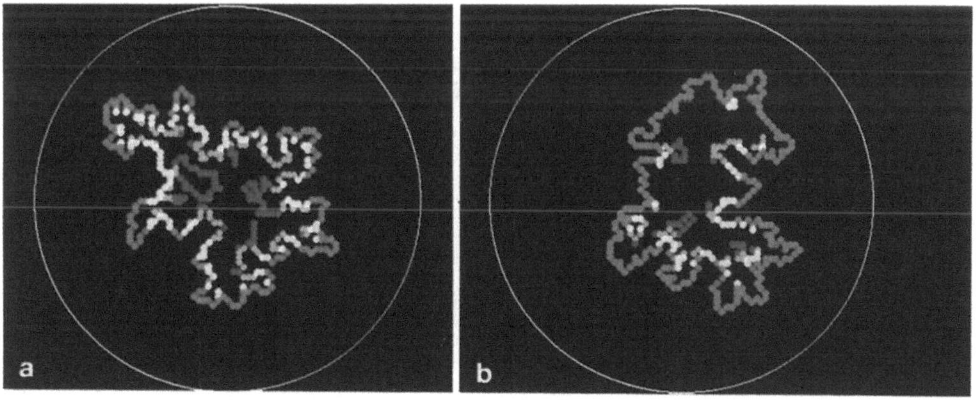

Fig.3 The growth probability distribution at perimeter sites of the same clusters as in Fig.2. The colors indicate the magnitude of growth probability which is equal to the concentration at that site, C(x): $0 \sim 1.0 \times 10^{-4}$, red; $\sim 3.0 \times 10^{-4}$, purple; $\sim$0.003, yellow; $\sim$0.02, green; and $\sim$0.02, sky blue, respectively.

Part VIII

Chemical Reactions and Hydrodynamic Instabilities

Global Mass Transport by Chemical Wave Propagation

H. Miike[1], *H. Hashimoto*[2], *and S.C. Müller*[2]

[1]Department of Electrical Engineering, Yamaguchi University, Tokiwadai 2557, Ube 755, Japan

[2]Max-Planck-Institut für Ernährungsphysiologie, Rheinlanddamm 201, D-4600 Dortmund 1, Fed. Rep. of Germany

ABSTRACT: Hydrodynamic flows accompanying chemical wave propagation are investigated in the Belousov-Zhabotinskii (BZ) reaction. A traveling cellular structure of the flow propagating with circular waves is detected even under a covered liquid/gas interface.

1. INTRODUCTION: Chemical wave propagation in the BZ reaction has been treated theoretically by models based on reaction-diffusion equations [1]. A variety of additional structures have been observed, however, which can be explained only by hydrodynamic effects. Examples are stationary "mosaic" patterns, transient "droplet"-like structures, or distortions and irregular decomposition of wave fronts [2,3]. Recently, we have carried out direct and quantitative measurements with respect to the role of hydrodynamic flows in such patterns. We used two-dimensional velocimetry and spectrophotometry based on microscope video imaging techniques and discovered that, in fact, chemical propagation is frequently accompanied by hydrodynamic flow. Two remarkable phenomena were detected even when the liquid/gas interface was covered, thus excluding evaporation: 1) a global flow structure traveling with circular waves in a chemically well reduced, excitable BZ solution layer, 2) oscillatory hydrodynamic flow under the condition of periodic passage of wave trains emanating from a rotating spiral center [4,5].

Here, we focus on the traveling flow structure accompanied by circular wave propagation. The geometric structure of the global flow and the possible origin of the flow are discussed.

2. MATERIALS AND METHODS : Samples of an excitable BZ solution catalyzed by ferroin were prepared as previously [3]. A volume of the mixture resulting in a 0.85 mm layer was placed in a Petri dish (diameter, 7 cm) and circular waves with almost constant period were triggered autonomously or by external stimuli at one or two locally stable points at the dish edges. Polystyrene particles (diameter, 0.48 μm) serving as scattering centers were added to the solution. Flow visualization was realized by illumination with a He-Ne laser beam under an inverse microscope. The velocity of the flow was analyzed by a newly developed dynamic image processing system consisting of a TV camera, an image processing board, and a PC & Convex vector computer network.

3. RESULTS AND DISCUSSION: In Fig.1, time traces of flow velocity are shown which were measured close to the surface (a) and the bottom (b) of the BZ solution layer at the center of the Petri dish. Near the surface, the speed of the flow increases as the wave front approaches the observation area. After wave passage, the flow suddenly reverses its direction (arrows) and then proceeds parallel to wave propagation. The hydrodynamic flow near the bottom shows the inverse behavior. Similar experiments were performed at

Springer Proceedings in Physics, Vol. 52 Dynamics and Patterns in Complex Fluids
Editors: A. Onuki · K. Kawasaki

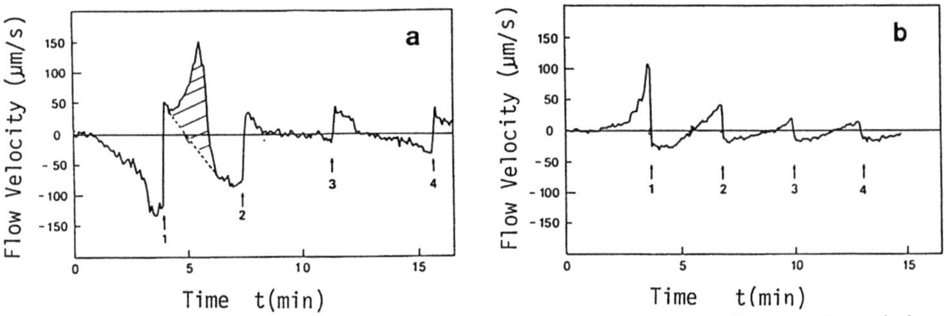

Fig.1 Time traces of flow velocity measured near the surface(a) and near the bottom (b).

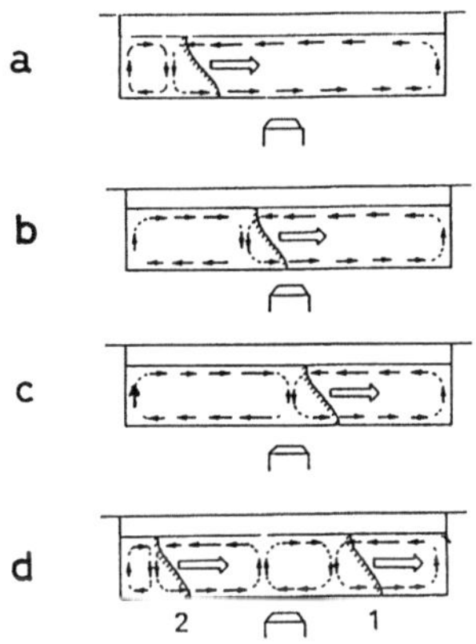

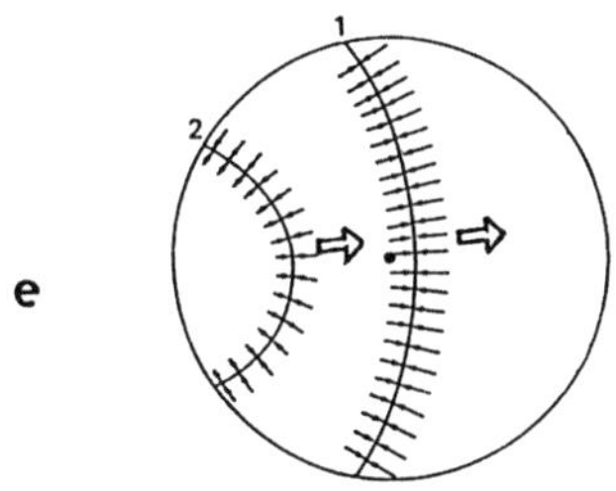

Fig.2 Schematic profiles of waves and hydrodynamic flow field. (a)-(d):side view; (e):top view.

different locations in the dish. Schematic profiles of wave fronts and the hydrodynamic flow field, as derived from these experiments, are shown in Fig.2, suggesting that an interface of a global cellular structure of the hydrodynamic flow travels with the propagating circular wave fronts. We tentatively call this structure a "traveling cell". The flow in this cell resembles the mean flow due to a concentration gradient (advection) observed in a simple liquid heated from the side. The concentration gradient of the catalyst ferroin at the wave front may play an important role in establishing the traveling cell.

The origin of the traveling cell is not clear at the present stage. For an explanation, not only the advection but also the dependence of surface tension on chemical composition and possibly reaction heat (Marangoni effect) may have to be taken into account.

Supported by the Stiftung Volkswagenwerk, Hannover. Discussions with Prof. B. Hess, Dr. Th. Plesser and Dr. H. Brand, and technical assistance of Mr. U. Heideck, Ms. G. Schulte and Ms. E. Schluter are acknowledged.

References

1)C.Vidal,P.Hanusse:Int.Rev.Phys.Chem.,5(1986)1.
2)K.I.Agladze,V.I.Krinsky,A.M.Pertsov:Nature 308(1984)834.
3)S.C.Müller,Th.Plesser,B.Hess:Naturwissenschaften 73(1986)165.
4)H.Miike,S.C.Müller,B.Hess:Chem.Phys.Lett.,114(1988)515.
5)H.Miike,S.C.Müller,B.Hess:Phys.Rev.Lett.,61(1988)2109.

Self-Collapse of the Target Pattern due to Macroscopic Fluctuation in the Belousov-Zhabotinsky Reaction

K. Suzuki

Department of Physics, Science University of Tokyo,
Kagurazaka, 1-3 Shinjuku, Tokyo, Japan

The stationary pattern of the Belousov-Zhabotinsky (B-Z) reaction has long been studied as a typical example of dissipative structures appearing in nonlinear systems far from equilibrium [1,2]. Recently, the chaotic behavior of the spiral pattern in the B-Z reaction has been extensively studied from the viewpoint of an order–disorder transition [3]. In the present article, we report several results of our observation of the dynamics of a sufficiently developed target pattern of the B-Z reaction.

The experiments were carried out in an unstirred B-Z reactive solution in a 150 mm diameter Petri dish, mounted in a water bath kept at 25°C. The ingredients of the solution were as follows: cerium ions 0.9 mM [as $Ce_2(SO_4)_3$], malonic acid 0.245 M, sulfuric acid 0.168 M, ferroin 0.52 mM, and bromine 0.3–0.69 M (as $NaBrO_3$). After these had been mixed, the solution was adequately agitated by a magnetic stirrer. Under these conditions, the medium soon generates oscillations, having periods from 20 to 50 s. After this, as soon as the solution has been moved into a sufficiently washed Petri dish with a fragment of filter paper on its bottom, a target pattern with wavelength 0.7–2 mm develops with the filter paper playing the role of pacemaker. The time evolution of the target pattern was measured with a CCD-TV camera, a 256 kbyte frame memory and a microcomputer.

A typical example of the self-collapse of the target pattern due to macroscopic fluctuation is shown in Fig. 1. Such a collapse of the target pattern never occurs when the initial bromine concentration is below 0.35 M: the pattern becomes gradually uni-

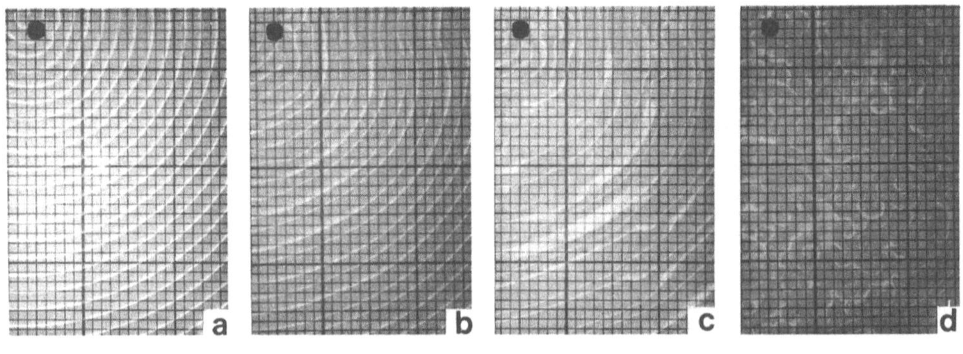

Fig. 1a–d. The evolution of the target pattern (initial bromine concentration 0.5 M). Stable growth (**a**), appearance of fluctuation (**b**), transition due to enhanced fluctuation (**c**), and collapse to a random pattern (**d**)

Springer Proceedings in Physics, Vol. 52 **Dynamics and Patterns in Complex Fluids**
Editors: A. Onuki · K. Kawasaki

form and the medium reaches thermal equilibrium after 5 h. When the initial bromine concentration is 0.5 M, a macroscopic fluctuation appears after 140 min, measured from the starting time of the reaction, and the pattern breaks down completely after 10 min, yielding a random pattern. The time required to generate macroscopic fluctuations decreases monotonically with increasing initial bromine concentration. In order to clarify the collapse process more quantitatively, the periods of the oscillating medium were measured at a point 15 mm away from the center, in the following two cases.

1) The target pattern is generated in reactive solutions with a pacemaker.
2) Bulk oscillation is generated in a clean Petri dish without a pacemaker. In this case the target pattern cannot be generated.

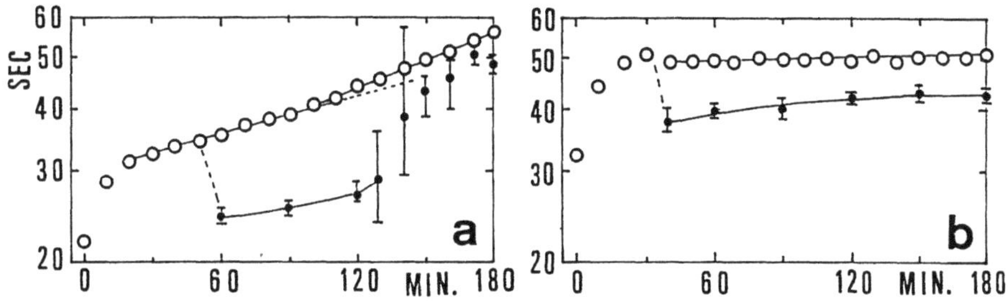

Fig. 2a,b. Variation of the periods of bulk oscillation (o) and target pattern (•) with time. The self-collapse process is clearly seen for an initial bromine concentration of 0.5 M (**a**), but not for an initial concentration of 0.35 M (**b**)

Figure 2 shows the period of bulk and pattern oscillations at the observation point as a function of time. The initial bromine concentration is 0.5 M in the case of Fig. 2a. The time profile of the bulk period is approximated by the two straight lines. In contrast to this, the profile of the period of the target pattern shows a slight increase initially but starts to change rapidly after 90 min; a macroscopic fluctuation is generated around 130 min, and the pattern finally breaks down at 140 min. Figure 2b is the case of an initial bromine concentration of 0.35 M. The time profile of the bulk period can be approximated by a gently sloping straight line, and the time profile of the target pattern is also a straight line parallel to that of the bulk period. In this case, no macroscopic fluctuations are generated.

The collapse of the target pattern can be well controlled by the initial bromine concentration. It is not affected by external perturbation or thermal convection. However, it is not observed in the ferroin system, where cerium is completely replaced by ferroin in the same bromine concentration. Thus it is concluded that the self-collapse of the target pattern seems to be governed by internal factors in the B-Z reaction system with a high concentration of bromine in addition to the presence of two kinds of metallic ions. In the presence of macroscopic fluctuation, diffusion occurs in the direction contrary to the progress of the pattern, and the system can partially recover its symmetry. The random pattern after collapse diffuses isotropically, and the system never recovers its global order again.

References

1. A.T. Winfree: Sci. Am. **230**, 82 (1974)
2. Y. Kuramoto: *Chemical Oscillations, Waves, and Turbulence*, Springer Ser. Syn., Vol. 19 (Springer, Berlin, Heidelberg 1984)
3. K.I. Agladze, V.I. Krinsky, A.M. Pertsov: Nature **308**, 834 (1984)

Chemical Sensing by Use of Entrainment

S. Nakata[1], *K. Yoshikawa*[2], *and H. Kawakami*[3]

[1]Nara University of Education, Nara 630, Japan
[2]College of General Education, Nagoya University, Nagoya 464-01, Japan
[3]Faculty of Engineering, University of Tokushima, Tokushima 770, Japan

An electrochemical system used as an electrical element exhibits various kinds of nonlinear characteristics, e.g., its capacitance changes markedly with applied voltage. In this report we propose a new sensing method for detecting electrochemical nonlinearity. A Wien bridge oscillator based on the phenomenon of frequency entrainment is used for this purpose. Experimental results are compared with those of theoretical simulation.

Figure 1 shows a sensing circuit called a Wien bridge oscillator with a sinusoidal external voltage source V_{in}. An electrochemical cell containing the test solution is connected as an element of the oscillator. When a sinusoidal voltage (1 V, 0.9 Hz) is applied to the circuit, the frequency of free oscillation is pulled in with the external frequency. That is, a frequency entrainment occurs.

Figure 2 shows the entrained response curves for various taste compounds to 5 mM sodium oleate. It is apparent that the oscillations are characteristically entrained to the external frequency by adding various chemical stimuli which depend on their taste category. It is assumed that each taste compound affects the capacitance of the electrode in a different manner [1, 2]. The reproducibility of this method is excellent. The pattern of the entrained response curves affords us useful information about electrochemical systems.

Theoretical simulation has been carried out for the Wien bridge oscillator with the voltage-dependent capacitance: $\mathbf{C_s = 0.01} + V_s$. The circuit equations are as follows:

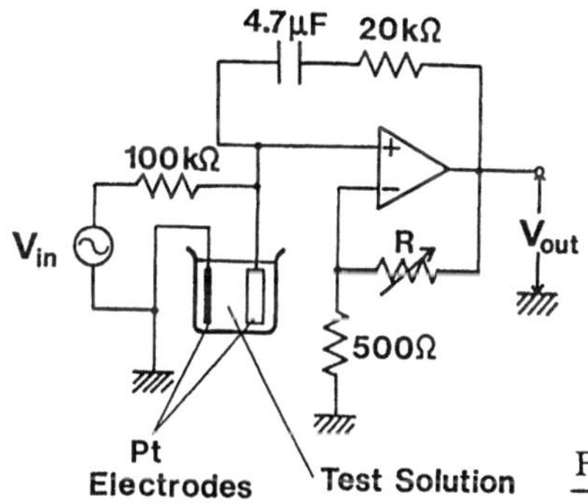

Fig. 1. Experimental apparatus.

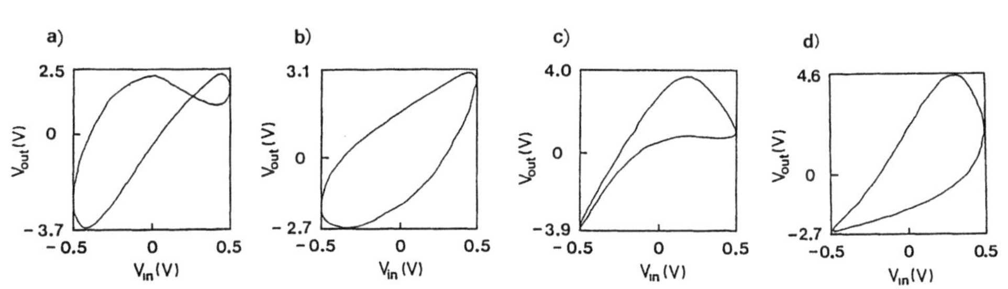

Fig. 2. The entrained response curves on adding chemical stimuli to aqueous solution of 5mM sodium oleate. a) 5mM sodium oleate, b) a) + 0.5mM KCl, c) a) + 0.3mM HCl, and d) a) + 0.5mM nicotine.

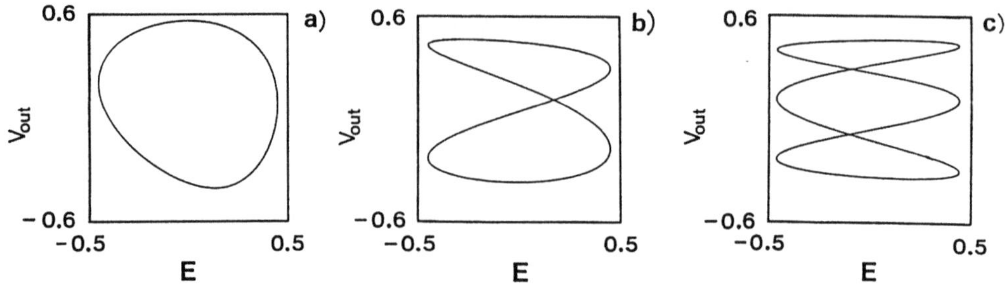

Fig. 3. Computer simulation of entrainments: a) E: $V_{out} = 1 : 1$, ($E_f = 5$), b) 2 : 1 ($E_f = 10$), and c) 3 : 1 ($E_f = 15$), where E(x 0.01) is the external voltage, V_{out} (x 0.01) the output voltage of the oscillation, and E_f the external frequency.

$$V_{out} = V_s + V_f + R_f C_f \, dV_f / dt \qquad (1)$$
$$C_f \, dV_f/dt = V_s / R_s + C_s \, dV_s / dt + (V_s - E(t)) / r \qquad (2)$$

Numerical results for Eqs. (1) and (2) are shown in Fig. 3(a), (b) and (c), which illustrate the entrained patterns: 1: 1, 2: 1 and 3: 1, respectively.

References

[1] K. Yoshikawa, M. Shoji, S. Nakata, S. Maeda, and H. Kawakami : Langmuir: **4**, 759-762 (1988)
[2] K. Yoshikawa, M. Shoji, and T. Ishii : BBRC: **160**, 699-704 (1989)

Nonlinear Response of Surface Waves in a Container Under an External Parametric Forcing

M. Umeki

Department of Physics, University of Tokyo, 7-3-1, Hongo, Bunkyo-ku, Tokyo 113, Japan

Abstract. Nonlinear response of two subharmonic modes of surface waves in a container subjected to vertical oscillation is studied based on the average Lagrangian method. A derived hamiltonian system with two degrees of freedom is shown to possess homoclinic chaos when suitable parameters are chosen. The dynamical system with a phenomenological damping term is considered and compared with recent experimental observations.

1. Introduction

The study of surface waves in a container which is forced to oscillate vertically goes back to Faraday's experiment (1831) [1]. Recent experiments [2-4,8,9] reveal that mode-competition occurs periodically or non-periodically in a certain region of two parameters, i.e., amplitude and frequency of the forcing $a_0 \cos 2\omega t$. A theoretical analysis based on an averaging method [5-7] leads to a hamiltonian system for slowly varying amplitudes p_i and q_i of each mode with a hamiltonian function:

$$\begin{aligned} H =& \frac{1}{2}\sum_{i=1}^{2}[A_0(p_i^2 - q_i^2) + \beta_i(p_i^2 + q_i^2) + \frac{A_i}{2}(p_i^2 + q_i^2)^2] \\ &+ \frac{C}{2}(p_1^2 + q_1^2)(p_2^2 + q_2^2) + \frac{D}{2}(p_1 q_2 - p_2 q_1)^2, \end{aligned} \tag{1}$$

where the normalized frequency offset to the i-th natural frequency and the forcing amplitude are denoted by $\beta_i = (\omega^2 - \omega_i^2)/(2\epsilon^2\omega^2)$ and $A_0 = a_0/a_1\epsilon^2$, respectively. The surface displacement $\eta(t,x,y)$ ((x,y) being the horizontal coordinates) is represented by

$$\eta(t,x,y) = \sum_i \epsilon a_i[p_i(\tau)\cos\omega t + q_i(\tau)\sin\omega t + O(\epsilon)]\psi_i(x,y),$$

where $a_i = (\kappa_i \tanh \kappa_i d)^{-1}$, $\tau = \epsilon^2\omega t$, κ_i is a wavenumber, d the depth, ψ_i an eigenfunction, and ϵ an expansion parameter. The hamiltonian system with a linear damping and analytically obtained nonlinear coefficients A_1, A_2, C, D yields a phase diagram including a region of attractors whose evolution appears non-periodic near the *codimension 2* bifurcation point, which is in good agreement with a recent experimental result [3].

2. Hamiltonian chaos in average surface-wave system

The hamiltonian system is integrable if the external forcing A_0 vanishes since two conservative quantities H and $\sum_{i=1}^{2}(p_i^2 + q_i^2)$ exist. Using the method of reduction

[10], we show that this system can be reduced to a single-degree-of-freedom hamiltonian system with 2π-periodic forcing. After the canonical transforms $(q_i, p_i) = \sqrt{2P_i'}(\cos Q_i', \ \sin Q_i')$, and $Q_1' = Q_1 + Q_2$, $Q_2' = Q_2$, $P_2' = P_2 - P_1$, $P_1' = P_1$, we obtain the reduced system :

$$\begin{aligned}
\dot{P}_1 &= -2DP_1(P_2 - P_1)\sin 2Q_1 - 2A_0P_1\sin 2(Q_1 + Q_2) - 2\alpha P_1,\\
\dot{Q}_1 &= \beta_1 - \beta_2 + EP_1 + FP_2 - D(P_2 - 2P_1)\cos 2Q_1 - A_0[\cos 2(Q_1 + Q_2) - \cos 2Q_2],\\
\dot{P}_2 &= -2A_0[P_1 \sin 2(Q_1 + Q_2) + (P_2 - P_1)\sin 2Q_2] - 2\alpha P_2, \qquad (2)\\
\dot{Q}_2 &= \beta_2 + 2A_2P_2 + FP_1 - DP_1\cos 2Q_1 - A_0\cos 2Q_2,\\
E &= 2(A_1 + A_2 - 2C - D), \quad F = 2(C - A_2) + D.
\end{aligned}$$

Using the Melnikov's function, we show that, under some assumptions and a sufficiently small forcing A_0, the unstable and stable manifolds of a hyperbolic saddle point intersect transversally and there exists a Smale horseshoe mapping in the neighbourhood of homoclinic orbits in $(X, Y) = \sqrt{2P_1}(\cos Q_1, \sin Q_1)$ plane [6].

A numerically obtained Poincaré section is shown in figure 1. Here the parameters are chosen as $(A_1, A_2, C, D, \beta_1, \beta_2, \alpha, P_2(0), A_0) =$ (-3,-3,-3,-4,0,-1,0,0.25,0.2) [case (a)] and (2,2,1,2,0,0,0,1,0.2) [case (b)]. The relations $A_1 = A_2$ and $\beta_1 = \beta_2$ come from the square symmetry and degeneracy of two modes in a square container. It should be noted that complete degeneracy yields no homoclinic orbits and there are heteroclinic orbits in case (b).

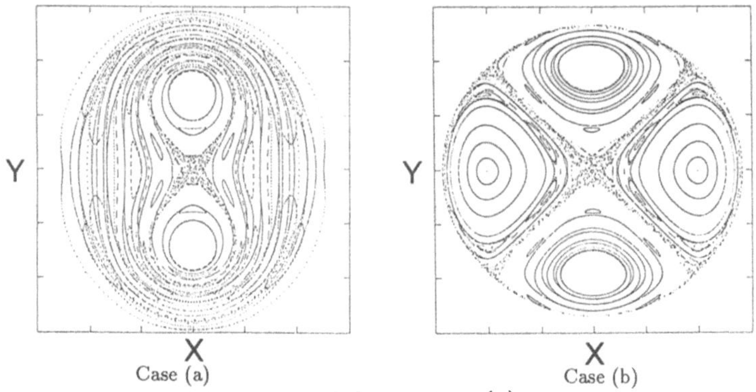

Figure 1. Poincaré section of the system (2)

References

[1] M. Faraday, Phil. Trans. R. Soc. Lond. **121**, 319 (1831).
[2] S. Ciliberto and J. P. Gollub, J. Fluid Mech. **158**, 381 (1985).
[3] F. Simonelli and J. P. Gollub, J. Fluid Mech. **199**, 471 (1989).
[4] M. Karatsu, Master thesis, Univ.of Tokyo. (1988) *in Japanese.*
[5] M. Umeki and T. Kambe, J. Phys. Soc. Jpn. **58**, 140 (1989).
[6] M. Umeki, (1989) submitted to J. Fluid Mech.
[7] J. W. Miles, J. Fluid Mech. **146**, 285 (1984).
[8] E. Meron and I. Procaccia, Phys. Rev. A **34**, 3221 (1986).
[9] Z. C. Feng and P. R. Sethna, J. Fluid Mech. **199**, 495 (1989).
[10] P. Holmes, J. Fluid Mech. **162**, 365 (1986).

Uphill Diffusion and Chemical Layering in a Mineral-Fluid System

T. Nishiyama

Department of Geology, Kyushu University 33, Fukuoka 812, Japan

Chemical layering in metamorphic rocks is a good example of self-organization in geology. Local equilibrium between a mineral and an intergranular fluid generates a coupling of diffusion flows which can cause diffusion up the concentration gradient. Resultant oscillatory concentration profile will produce chemical layering, when associated with dissolution-precipitation reactions.

Consider a ternary system consisting of components 1 and 2, and H_2O as solvent. The coupling of diffusion flows in the system can cause diffusion up the concentration gradients, depending on the relative magnitude of L_{11}/L_{12} and L_{22}/L_{12}. The condition of equilibrium between the fluid (the medium for diffusion) and mineral M ($M = \nu_1\alpha + \nu_2\beta$) gives this kind of coupling, where ν_1 and ν_2 stand for stoichiometric coefficients, and α and β denotes fluid species of components 1 and 2, respectively. A formal expression for diffusion flow of component 1 can be written as $J_1 = -L_{11}(\delta\mu_1/\delta x) - L_{12}(\delta\mu_2/\delta x)$ in one dimension. The Gibbs-Duhem condition for mineral M gives the relation $\nu_1\delta\mu_1 + \nu_2\delta\mu_2 = 0$ at constant temperature and pressure. Combining these relations, we get

$$J_1 = -(L_{11} - (\nu_1/\nu_2)L_{12})(\delta\mu_1/\delta x). \qquad (1)$$

Hence the condition for uphill diffusion becomes

$$L_{11}/L_{12} < \nu_1/\nu_2. \qquad (2)$$

This relation together with Onsager's relation $L_{11}L_{22} > (L_{12})^2$ defines the region of uphill diffusion in the L-ratio diagram (a plot of L_{11}/L_{12} vs. L_{22}/L_{12}) as shown in Fig. 1. The concentration dependence of D_{ij}, $c_1/c_2 = D_{12}/D_{21}$, stems from Onsager's reciprocity relation, $L_{ij} = L_{ji}$ ($i \neq j$), with an assumption of ideal solution for the fluid. The functional form of $D_{ij}(c_1, c_2)$ is unknown, however, we can prove that there is always a critical concentration $(c_i)_{cr}$ for uphill diffusion, irrespective of the functional form of $D_{ij}(c_1, c_2)$. The magnitude of $(c_i)_{cr}$ depends on $D_{ij}(c_1, c_2)$. The relative magnitude of $(c_i)_{cr}$ and the mean concentration in the fluid (c_i) will determine the behaviour of the system. Fluctuations

Springer Proceedings in Physics, Vol. 52 **Dynamics and Patterns in Complex Fluids**
Editors: A. Onuki · K. Kawasaki

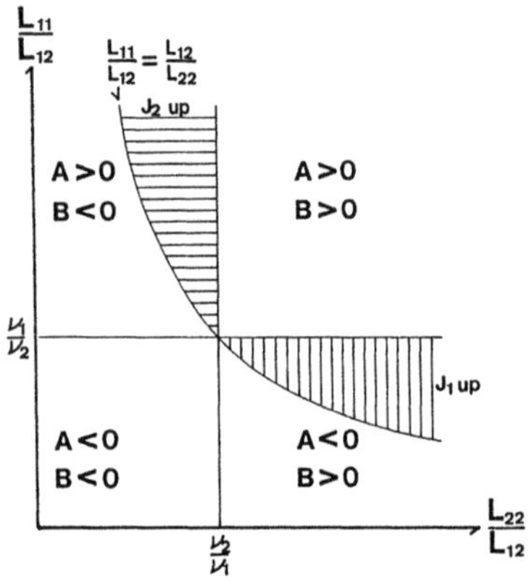

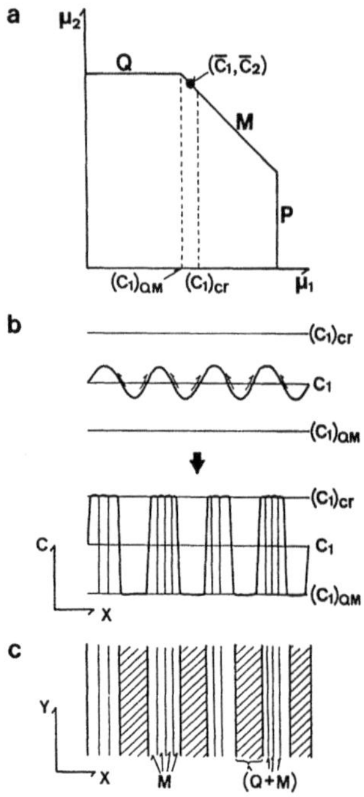

Fig.1(left). The L-ratio diagram for the model system. The positive quadrant of the diagram consists of four regions divided by two lines: $A=L_{11}-(v_1/v_2)L_{12}=0$ and $B=L_{22}-(v_2/v_1)L_{12}=0$. The signs of A and B define whether the diffusion flows, J_1 (A) and J_2 (B), are uphill (negative) or downhill (positive). The condition of $(L_{11}/L_{12})>(L_{12}/L_{22})$ stems from positive entropy production. The domains hatched with vertical and horizontal lines represent those in which J_1 and J_2, respectively, are uphill.

Fig.2(right). Qualitative behaviour of the system with dissolution-precipitation reactions. a. If the initial state is between the invariant point, $(c_1)_{QM}$, and the critical point, $(c_1)_{cr}$, fluctuations will grow by uphill diffusion. b. Positive fluctuations will reach $(c_1)_{cr}$, resulting in a flat top with sharp pulses. Negative fluctuations and sharp pulses will finally arrive at $(c_1)_{QM}$, where a part of M dissolves and Q precipitates. c. The formation of layering consisting of M and (M+Q) layers can be expected. M also involves the second order layering indicated by the presence of thin horizons of (Q+M).

in concentration will grow when $c_i < (c_i)_{cr}$, resulting in a periodic concentration profile (Fig.2). The concentration gradient of the other component will also grow by dissolution-precipitation reaction of mineral M due to the local equilibrium constraint, although the component shows downhill diffusion. This situation will continue until c_i reaches the invariant point where a new mineral Q will precipitate. Thus the metamorphic layering can be explained by uphill diffusion associated with dissolution-precipitation reactions.

Experiment on Magnetoadsorption of NO on Micropores

H. Uchiyama[1], *S. Ozeki*[1], *K. Kaneko*[1], *Y. Natsume*[2], *and T. Suzuki*[2]

[1]Department of Chemistry, Chiba University, Chiba 260, Japan
[2]Department of Physics, Chiba University, Chiba 260, Japan

Remarkable transient enhancement is observed in the adsorption of NO on micropores of micrographitic carbons by application of a magnetic field. The observed results are discussed in connection with the effect of an induction current.

1. Introduction

Remarkable phenomena in the adsorption of NO molecules on microporous carbons open new aspects [1,2] of the physics-chemistry interface. It is well confirmed [3] that the dimerization of NO is associated with noteworthy micropore filling of supercritical NO on surface-modified microporous carbons. Further, three of the present authors (H.U, S.O and K.K) have recently reported [4] that application of a magnetic field H provides a transient increase of the amount of NO adsorption on carbon fibers with slit-shaped micropores. In this paper, on the basis of experimental results, a model for the mechanism is proposed.

2. Experiment

The concept of the experiment is schematically shown in Fig.1. The amount of NO adsorption is determined volumetrically at B. The sensitivity of the adsorption detection is 2μg/g-adsorbent. The magnetic field H is applied by permanent magnets M. Measurements are made in quasi-adsorption equilibrium at 1.8 ± 0.2kPa, which is obtained 43-44 hours after the gas introduction.

We show a typical example of the enhancement of NO adsorption by black circles in Fig.2, where polyacrylonitrile(PAN)-based carbon fibers are used as the micropores at H=7.6kG and T=303.0 ± 0.2K. On the other hand, we can find no enhancement in the case of nonporous carbon black, as shown by white circles in Fig.2.

3. Proposal of a model

We propose the following dynamical model for the mechanism of the magnetoadsorption mentioned above, in connection with the reported fact [5,6] that the adsorption of H_2O molecules disposes the structure of micrographitic carbons to change significantly.

1. The induction current appears on the surface of micropores by the application of the field (dH/dt).
2. Such a local current enhances the lattice-vibration for some distorted modes.
3. This enhancement of the lattice-vibration causes the transition from the stable structure to the quasi-stable one as shown schematically from S to QS in Fig.3.

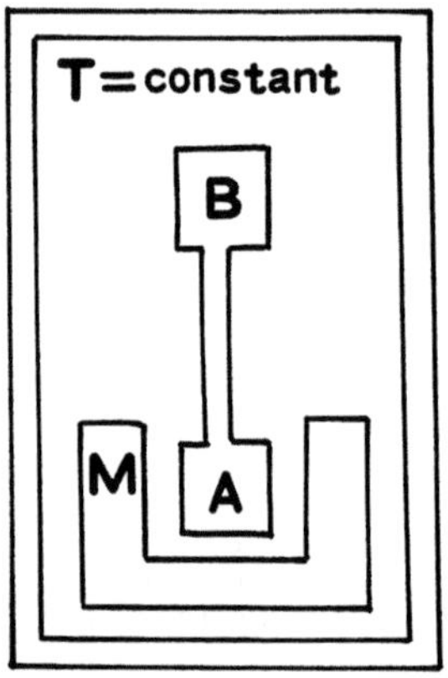

Fig.1

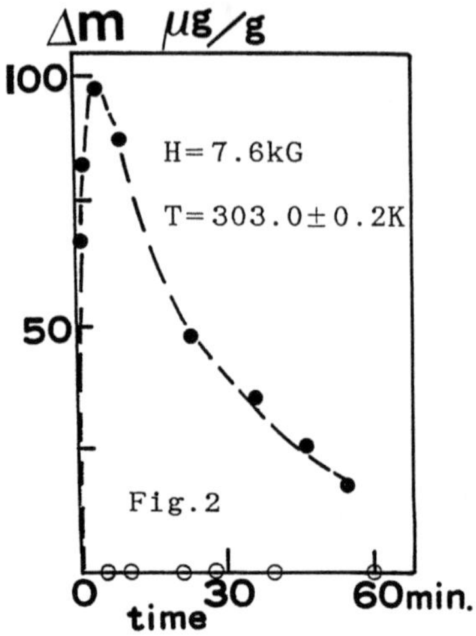

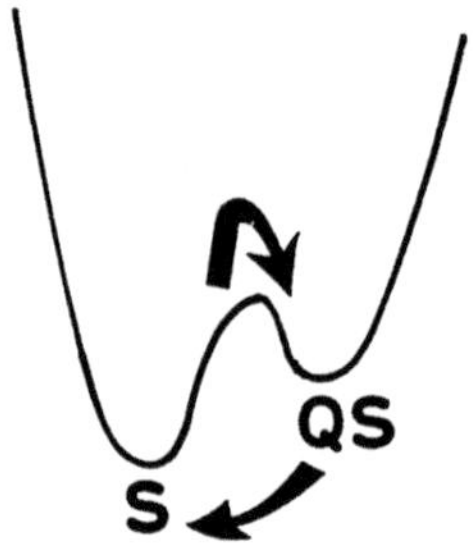

Fig.3

Fig.1 The concept of the experiment. The sample is placed at A and the pressure at B is measured in the tube for NO gas. The magnet M is used to apply the external field. Temperature is kept constant.

Fig.2 The observed transient enhancement of the amount of adsorption Δm by the field. The units of Δm are μg per gram of adsorbent.

Fig.3 Schematic of the proposed model: The structure change from the stable state S to the quasi-stable one QS is caused by the induction current.

4. The new quasi-stable structure (QS in Fig.3) has a larger capacity for the adsorption of NO{$2NO \rightarrow (NO)_2$} than the stable one S.
5. The thermal energy generated essentially in the process 1→2→3 is transferred to the heat bath because of the increase in entropy.
6. The new structure relaxes to the stable one. However, it takes a long relaxation time, about 10~100 min.

These processes are just assumptions and involve problems to be solved in future. Experimentally, the dependence of the magnetoadsorption on dH/dt should be measured as well as the dependence on H itself. For example, the observation in a rotating magnetic field is quite desirable. In addition, we should directly detect the amount of adsorption for micropores at A in Fig.1. Theoretically, the explanation for each process should be refined. In particular, we would like to express the structure change in process 3 in the deformation of the pattern for micrographitic carbons. In the phenomena of magnetoadsorption, many attractive problems are open to both physicists and chemists.

[1] K.Kaneko, N.Fukuzaki and S.Ozeki, J.Chem.Phys.87,776(1987).
[2] K.Kaneko, N.Fukuzaki, K.Kakei, T.Suzuki and S.Ozeki, Langmuir 5,960(1989).
[3] K.Kaneko, Colloid Surf.37,115(1989).
[4] H.Uchiyama, S.Ozeki and K.Kaneko, submitted to J.Chem.Phys.
[5] K.Kaneko, Y.Fujiwara and K.Nishikawa, J.Colloid Interface Science 127,298(1989).
[6] T.Suzuki and K.Kaneko, Carbon 26,743(1988).

Index of Contributors

MIX
Papier aus verantwortungsvollen Quellen
Paper from responsible sources
FSC® C105338

If you have any concerns about our products,
you can contact us on
ProductSafety@springernature.com

In case Publisher is established outside the EU,
the EU authorized representative is:
Springer Nature Customer Service Center GmbH
Europaplatz 3, 69115 Heidelberg, Germany

Printed by Libri Plureos GmbH
in Hamburg, Germany